Recent Advances in Gastrointestinal Cancers: From Microbiota Modulation to New Therapeutic Approaches

Recent Advances in Gastrointestinal Cancers: From Microbiota Modulation to New Therapeutic Approaches

Guest Editors

Serena Martinelli
Elena Niccolai

MDPI

Basel • Beijing • Wuhan • Barcelona • Belgrade • Novi Sad • Cluj • Manchester

Guest Editors
Serena Martinelli
Department of Experimental
and Clinical Medicine
University of Florence
Florence
Italy

Elena Niccolai
Department of Experimental
and Clinical Medicine
University of Florence
Florence
Italy

Editorial Office
MDPI AG
Grosspeteranlage 5
4052 Basel, Switzerland

This is a reprint of the Special Issue, published open access by the journal *Biomedicines* (ISSN 2227-9059), freely accessible at: https://www.mdpi.com/journal/biomedicines/special_issues/Gastrointestinal_Cancers.

For citation purposes, cite each article independently as indicated on the article page online and as indicated below:

Lastname, A.A.; Lastname, B.B. Article Title. *Journal Name* **Year**, *Volume Number*, Page Range.

ISBN 978-3-7258-5769-2 (Hbk)
ISBN 978-3-7258-5770-8 (PDF)
https://doi.org/10.3390/books978-3-7258-5770-8

Contents

biomedicines

Editorial

Editorial: Recent Advances in Gastrointestinal Cancers: From Microbiota Modulation to New Therapeutic Approaches

Serena Martinelli * and Elena Niccolai *

Department of Clinical and Experimental Medicine, University of Florence, 50139 Florence, Italy
* Correspondence: serena.martinelli@unifi.it (S.M.); elena.niccolai@unifi.it (E.N.)

Gastrointestinal (GI) cancers, including colorectal, pancreatic, and biliary tract malignancies, represent a major burden worldwide, characterized by high incidence, mortality, and clinical heterogeneity [1,2]. Despite advances in early detection and standard therapies, many patients present with refractory or aggressive disease, underscoring the urgent need for innovative therapeutic strategies, precise prognostic tools, and preclinical models that faithfully recapitulate human tumor biology.

The contributions included in this Special Issue reflect the multidimensional progress in the field, ranging from pharmacological innovations and biomarker discovery to patient-derived models and immune regulation, all framed within the evolving paradigm of precision oncology [3].

Advances in pharmacological management remain at the forefront. Regorafenib in combination with 5-fluorouracil (5-FU) has shown promising disease control and acceptable safety in heavily pretreated metastatic colorectal cancer (mCRC) [4]. Complementing this, real-world data suggest that the addition of bevacizumab to trifluridine–tipiracil (FTD-TPI) enhances progression-free survival and disease control rates, reinforcing the value of combinatorial strategies in mCRC clinical practice [5]. The efficacy of precision oncology relies also on robust molecular biomarkers for risk stratification and therapy guidance [6].

Similarly, multi-omics analyses in colon adenocarcinoma (COAD) identified hub genes, including Claudin1 (*CLDN1*), inhibin subunit beta A (*INHBA*), and chemokine (C-X-C motif) ligand 12 (*CXCL12*), as potential prognostic biomarkers, with single-cell RNA sequencing revealing cell-type-specific expression patterns [7]. In parallel, investigations into the Wnt/β-catenin pathway in colorectal adenomas underscore the importance of early molecular events and their translational relevance as biomarkers and therapeutic targets [8]. Together, these studies emphasize the integration of genomic, transcriptomic, and epigenetic data to advance precision medicine in GI oncology. Alongside these pharmacological and molecular advances, patient-derived organoids have become powerful platforms to evaluate therapeutic responses and interrogate disease mechanisms [9]. Organoids are three-dimensional, organ-like structures generated from self-organizing stem cells. They display organ-specific features and arise from stem cells undergoing intrinsic self-organization. Compared to traditional two-dimensional cell cultures, organoids offer significant advantages as they more closely replicate physiological cellular composition and function [10]. Organoid models derived from primary sclerosing cholangitis (PSC) and cholangiocarcinoma made possible the testing of JAK inhibitors and chemotherapy regimens while simultaneously exploring STAT3 expression in tumor and immune compartments [11]. Such models bridge the gap between molecular discoveries and clinical applications, providing a physiologically relevant context to predict patient specific responses [12]. Although not directly covered by the papers in this Special

 https://doi.org/10.3390/biomedicines13102540

Issue, two emerging directions deserve attention for their potential to shape the future of GI oncology. The first is surgical innovation, particularly the development of intraoperative fluorescence-guided techniques. Indocyanine green (ICG) remains the most commonly used fluorophore in clinical practice for visualizing lymphatic drainage and facilitating sentinel lymph node mapping, yet its lack of tumor specificity represents a major limitation to its broader oncologic utility [13,14]. This limitation underscores the need for next-generation fluorescent contrast agents that selectively target tumor-specific surface biomarkers, thereby enabling more accurate intraoperative discrimination between malignant and healthy tissue [15]. Ongoing research is actively addressing this gap, aiming to develop targeted probes capable of enhancing both surgical precision and oncologic outcomes [9,16,17]. The second frontier, perhaps even more transformative, is the modulation of the gut microbiota. Altered microbial ecosystems are increasingly recognized as key players in colorectal carcinogenesis and tumor immunity. Recent studies have shown that microbiome profiling can discriminate between adenomatous polyps and colorectal cancer, highlighting microbial signatures as potential diagnostic and prognostic tools [18]. In parallel, advanced in vitro and in silico platforms are being developed to model the interactions between microbial communities, immune cells, and tumors, providing unprecedented opportunities to unravel mechanisms of immune modulation and treatment response [19]. The concept of immunonutrition, integrating diet, probiotics, and prebiotics into oncology care, is also gaining traction as a strategy to restore eubiosis, enhance immune surveillance, and synergize with pharmacological and immunotherapeutic interventions [20], and compelling evidence indicates that the gut microbiome shapes the efficacy and toxicity of chemotherapy and immunotherapy in GI cancers [21,22]. Approaches such as probiotics, prebiotics, dietary strategies, and even fecal microbiota transplantation are being actively investigated as adjunct therapies capable of overcoming resistance and potentiating antitumor immunity [23–25]. Taken together, the contributions in this Special Issue underscore the progress being made in GI oncology, particularly in pharmacological strategies, biomarker development, organoid technologies and immune regulation. At the same time, the exploration of surgical innovations and microbiota modulation exemplifies the dynamic expansion of the field toward a more holistic and personalized model of care. By integrating these avenues, precision oncology in GI cancers has the potential to significantly improve patient outcomes in the years to come.

Author Contributions: Conceptualization, S.M. and E.N.; methodology, S.M. and E.N.; writing—original draft preparation, S.M. and E.N.; writing—review and editing, S.M. and E.N.; visualization, S.M. and E.N.; supervision, S.M.; project administration, S.M. and E.N. All authors have read and agreed to the published version of the manuscript.

Conflicts of Interest: The authors declare no conflicts of interest.

References

1. Bray, F.; Laversanne, M.; Sung, H.; Ferlay, J.; Siegel, R.L.; Soerjomataram, I.; Jemal, A. Global cancer statistics 2022: GLOBOCAN estimates of incidence and mortality worldwide for 36 cancers in 185 countries. *CA Cancer J. Clin.* **2024**, *74*, 229–263. [CrossRef]
2. Elmadani, M.; Mokaya, P.O.; Omer, A.A.A.; Kiptulon, E.K.; Klara, S.; Orsolya, M. Cancer burden in Europe: A systematic analysis of the GLOBOCAN database (2022). *BMC Cancer* **2025**, *25*, 447. [CrossRef]
3. Alsina, M.; Huerta, A.E.; Lordick, F.; Verschueren, S.; Moehler, M.; Fontana, E.; Smyth, E.; Sclafani, F.; Wagner, A.D.; Rimassa, L.; et al. Current practices and challenges in implementing precision medicine for upper gastrointestinal cancers in European academic centers: An EORTC survey. *ESMO Gastrointest. Oncol.* **2024**, *5*, 100074. [CrossRef]
4. Abdelrahim, M.; Esmail, A.; Al-Najjar, E.; Khasawneh, B.; Umoru, G.; Abdelrahim, W.; Abboud, K.; Ajewole, V.B. Safety and Efficacy of Regorafenib and 5-Fluorouracil Combination Therapy in Refractory Metastatic Colorectal Cancer After Third-Line Treatment: An Institutional Experience. *Biomedicines* **2025**, *13*, 1151. [CrossRef]

5. Kim, H.; Shin, K.; An, H.J.; Kim, I.-H.; Bae, J.H.; Lee, Y.S.; Lee, I.K.; Lee, M.; Park, S.J. Real-World Comparison of Trifluridine–Tipiracil with or Without Bevacizumab in Patients with Refractory Metastatic Colorectal Cancer. *Biomedicines* **2025**, *13*, 976. [CrossRef] [PubMed]
6. Dang, D.K.; Park, B.H. Circulating tumor DNA: Current challenges for clinical utility. *J. Clin. Investig.* **2022**, *132*, e154941. [CrossRef]
7. Cheon, J.; Kim, S.H.; Park, J.; Kim, T.H. Prognostic Significance of CLDN1, INHBA, and CXCL12 in Colon Adenocarcinoma: A Multi-Omics and Single-Cell Approach. *Biomedicines* **2025**, *13*, 1035. [CrossRef]
8. Tufail, M.; Jiang, C.H.; Li, N. Wnt signaling in cancer: From biomarkers to targeted therapies and clinical translation. *Mol. Cancer* **2025**, *24*, 107. [CrossRef] [PubMed]
9. Martinelli, S.; Peri, S.; Anceschi, C.; Laurenzana, A.; Fortuna, L.; Mello, T.; Naldi, L.; Marroncini, G.; Tricomi, J.; Biagioni, A.; et al. Targeting CEACAM5: Biomarker Characterization and Fluorescent Probe Labeling for Image-Guided Gastric Cancer Surgery. *Biomedicines* **2025**, *13*, 1812. [CrossRef] [PubMed]
10. Park, G.; Rim, Y.A.; Sohn, Y.; Nam, Y.; Ju, J.H. Replacing Animal Testing with Stem Cell-Organoids: Advantages and Limitations. *Stem. Cell Rev. Rep.* **2024**, *20*, 1375–1386. [CrossRef]
11. Boden, C.; Esser, L.K.; Dold, L.; Langhans, B.; Zhou, T.; Kaczmarek, D.J.; Gonzalez-Carmona, M.A.; Weismüller, T.J.; Kristiansen, G.; Kalff, J.C.; et al. The IL-6/JAK/STAT3 Axis in Cholangiocarcinoma and Primary Sclerosing Cholangitis: Unlocking Therapeutic Strategies Through Patient-Derived Organoids. *Biomedicines* **2025**, *13*, 1083. [CrossRef]
12. Cox, K.E.; Turner, M.A.; Lwin, T.M.; Amirfakhri, S.; Kelly, K.J.; Hosseini, M.; Ghosh, P.; Obonyo, M.; Hoffman, R.M.; Yazaki, P.J.; et al. Targeting Patient-Derived Orthotopic Gastric Cancers with a Fluorescent Humanized Anti-CEA Antibody. *Ann. Surg. Oncol.* **2024**, *31*, 6291–6299. [CrossRef]
13. Baldari, L.; Boni, L.; Cassinotti, E. Lymph node mapping with ICG near-infrared fluorescence imaging: Technique and results. *Minim. Invasive Ther. Allied Technol.* **2023**, *32*, 213–221. [CrossRef]
14. Liao, Y.; Zhao, J.; Chen, Y.; Zhao, B.; Fang, Y.; Wang, F.; Wei, C.; Ma, Y.; Ji, H.; Wang, D.; et al. Mapping Lymph Node during Indocyanine Green Fluorescence-Imaging Guided Gastric Oncologic Surgery: Current Applications and Future Directions. *Cancers* **2022**, *14*, 5143. [CrossRef]
15. Martinelli, S.; Fortuna, L.; Coratti, F.; Passagnoli, F.; Amedei, A.; Cianchi, F. Potential Probes for Targeted Intraoperative Fluorescence Imaging in Gastric Cancer. *Cancers* **2024**, *16*, 4141. [CrossRef] [PubMed]
16. Gutowski, M.; Framery, B.; Boonstra, M.C.; Garambois, V.; Quenet, F.; Dumas, K.; Scherninski, F.; Cailler, F.; Vahrmeijer, A.L.; Pèlegrin, A. SGM-101: An innovative near-infrared dye-antibody conjugate that targets CEA for fluorescence-guided surgery. *Surg. Oncol.* **2017**, *26*, 153–162. [CrossRef]
17. Cox, K.E.; Turner, M.A.; Amirfakhri, S.; Lwin, T.M.; Hosseini, M.; Ghosh, P.; Obonyo, M.; Murakami, T.; Hoffman, R.M.; Yazaki, P.J.; et al. Humanized Anti-Carcinoembryonic Antigen Antibodies Brightly Target and Label Gastric Cancer in Orthotopic Mouse Models. *J. Surg. Res.* **2024**, *293*, 701–708. [CrossRef] [PubMed]
18. Rotelli, A.; Salman, A.; Di Gloria, L.; Nannini, G.; Niccolai, E.; Luschi, A.; Amedei, A.; Iadanza, E. Analysis of Microbiome for AP and CRC Discrimination. *Bioengineering* **2025**, *12*, 713. [CrossRef] [PubMed]
19. Bertorello, S.; Cei, F.; Fink, D.; Niccolai, E.; Amedei, A. The Future Exploring of Gut Microbiome-Immunity Interactions: From In Vivo/Vitro Models to In Silico Innovations. *Microorganisms* **2024**, *12*, 1828. [CrossRef]
20. Martinelli, S.; Lamminpää, I.; Dübüş, E.N.; Sarıkaya, D.; Niccolai, E. Synergistic Strategies for Gastrointestinal Cancer Care: Unveiling the Benefits of Immunonutrition and Microbiota Modulation. *Nutrients* **2023**, *15*, 4408. [CrossRef]
21. Kolahi Sadeghi, L.; Vahidian, F.; Eterafi, M.; Safarzadeh, E. Gastrointestinal cancer resistance to treatment: The role of microbiota. *Infect. Agents Cancer* **2024**, *19*, 50. [CrossRef]
22. Chrysostomou, D.; Roberts, L.A.; Marchesi, J.R.; Kinross, J.M. Gut Microbiota Modulation of Efficacy and Toxicity of Cancer Chemotherapy and Immunotherapy. *Gastroenterology* **2023**, *164*, 198–213. [CrossRef]
23. Helmink, B.A.; Khan, M.A.W.; Hermann, A.; Gopalakrishnan, V.; Wargo, J.A. The microbiome, cancer, and cancer therapy. *Nat. Med.* **2019**, *25*, 377–388. [CrossRef] [PubMed]
24. Matson, V.; Fessler, J.; Bao, R.; Chongsuwat, T.; Zha, Y.; Alegre, M.L.; Luke, J.J.; Gajewski, T.F. The commensal microbiome is associated with anti-PD-1 efficacy in metastatic melanoma patients. *Science* **2018**, *359*, 104–108. [CrossRef] [PubMed]
25. Almonte, A.A.; Thomas, S.; Zitvogel, L. Microbiota-centered interventions to boost immune checkpoint blockade therapies. *J. Exp. Med.* **2025**, *222*, e20250378. [CrossRef] [PubMed]

biomedicines

MDPI

Article

The Interplay among Wnt/β-catenin Family Members in Colorectal Adenomas and Surrounding Tissues

Domenica Lucia D'Antonio [1,2], Fabiana Fantini [1], Carmelo Moscatello [1], Alessio Ferrone [1], Stefano Scaringi [3], Rosa Valanzano [3], Ferdinando Ficari [3], Konstantinos Efthymakis [4], Matteo Neri [4], Gitana Maria Aceto [1] and Maria Cristina Curia [1,*]

1 Department of Medical, Oral and Biotechnological Sciences, "Gabriele d'Annunzio" University of Chieti-Pescara, 66100 Chieti, Italy; domenica.dantonio@unich.it (D.L.D.); fanfaby@gmail.com (F.F.); carmelo.moscatello@unich.it (C.M.); alessioferrone@yahoo.it (A.F.); gitana.aceto@unich.it (G.M.A.)

2 Villa Serena Foundation for Research, Via Leonardo Petruzzi 42, 65013 Città Sant'Angelo, Italy

3 Department of Clinical and Experimental Medicine, University of Florence, Largo Brambilla 3, 50134 Firenze, Italy; stefano.scaringi@unifi.it (S.S.); rosa.valanzano@unifi.it (R.V.); ferdinando.ficari@unifi.it (F.F.)

4 Department of Medicine and Aging Sciences, "Gabriele d'Annunzio" University of Chieti-Pescara, 66100 Chieti, Italy; efkn78@gmail.com (K.E.); mneri@unich.it (M.N.)

* Correspondence: mariacristina.curia@unich.it

Abstract: Background: The colorectal adenoma undergoes neoplastic progression via the normal epithelium–adenoma–adenocarcinoma sequence as reported in the Vogelgram. The hazard of developing a tumor is deeply associated with the number and size of adenomas and their subtype. Adenomatous polyps are histologically categorized as follows: approximately 80–90% are tubular, 5–15% are villous, and 5–10% are tubular/villous. Given the higher risk of a malignant transformation observed in tubular/villous adenomas, patients diagnosed with adenomatous polyposis are at an improved risk of developing CRC. The Wnt/β-catenin pathway plays a key role in the onset of colorectal adenoma; in particular, intestinal cells first acquire loss-of-function mutations in the *APC* gene that induce the formation of adenomas. Methods: Wnt/β-catenin pathway *APC, Wnt3a, Wnt5a, LEF1,* and *BCL9* genes and protein expression analyses were conducted by qRT-PCR and western blot in 68 colonic samples (polyps and adjacent mucosa) from 41 patients, of which 17 were affected by FAP. Ten normal colonic mucosal samples were collected from 10 healthy donors. Results: In this study, both the *APC* gene and protein were less expressed in the colon tumor compared to the adjacent colonic mucosa. Conversely, the activated β-catenin was more expressed in polyps than in the adjacent mucosa. All results confirmed the literature data on carcinomas. A statistically significant correlation between *Wnt3a* and *BCL9* both in polyps and in the adjacent mucosa underlines that the canonical Wnt pathway is activated in early colon carcinogenesis and that the adjacent mucosa is already altered. Conclusion: This is the first study analyzing the difference in expression of the Wnt/β-catenin pathway in human colorectal adenomas. Understanding the progression from adenomas to colorectal carcinomas is essential for the development of new therapeutic strategies and improving clinical outcomes with the use of APC and β-catenin as biomarkers.

Keywords: CRA; colorectal adenoma; *APC; Wnt3a; Wnt5a; LEF1; BCL9;* polyps; early carcinogenesis; Wnt/β-catenin

1. Introduction

The prevalence of colorectal adenomas (CRAs) increases with age, mainly in Western populations—30–40% in the people over 50 years, predominantly in men [1,2]. The annual rate of adenoma progression to colorectal cancer (CRC) is ~0.25% [3].

CRA is associated with CRC, and at least 80% of CRC undergoes neoplastic progression via the normal epithelium–adenoma–adenocarcinoma sequence as reported in the

Vogelgram model [4,5]. The incidence of cancer after a negative colonoscopy is significant because adenomas may be missed during a colonoscopy, or biological changes in the tumor growth rate may occur [6]. Screening and surveillance programs can help identify precursor lesions and prevent death from CRC [7]. Thus, it is important to understand the progression from CRA to carcinomas to facilitate the development of novel treatment strategies and improve clinical outcomes.

The malignancy of adenomas is highly correlated with the occurrence of colon cancer, depending on the subtype [8]. Furthermore, the risk of developing cancer is closely linked to the number and dimensions of previously identified polyps [9]. Developing multiple colonic polyps with malignant potential increases the lifetime risk of developing colorectal cancer (CRC). There are at least three types of polyps based on the histology and molecular pathway: adenomatous, serrated, or hyperplastic (non-neoplastic) [10–12]. The first type is characterized by the adenomatous histotype, while both sessile/traditional serrated adenomas and hyperplastic polyps have a serrated histotype [13]. The adenomatous histotype can be tubular (more than 80%), villous (5–15%), and tubular/villous (5–15%). Hyperplastic polyps are common and carry a small risk of evolving into cancer [8]. Although the different types of polyps may be diffused in the large bowel, adenomatous and hyperplastic polyps are mainly located in the distal colon [14–16] and sessile serrated polyps are frequently found in the proximal colon [17–20]. Owing to the malignant potential of tubular/villous adenomas, patients diagnosed with adenomatous polyposis, i.e., the constitutive development of multiple colorectal adenomas, are at an increased risk of developing CRC. Most studies investigating the carcinogenesis of *CRA* have focused on villous and familial adenomatous polyps, which have the highest rates of carcinogenesis [21,22], whereas few studies have investigated sporadic tubular adenomas, which has the highest clinical incidence [23,24]. Polyps serve as direct precursors to colorectal cancer in families with a history of polyposis syndrome, as well as in the general population. Genetic events, such as the gain or loss of function of molecules essential for intestinal cell homeostasis, may lead to the development of polyps [25].

The Wnt/β-catenin signaling deregulation is an early event in the onset of colorectal adenoma [26]. Its upregulation is mainly due to the altered functions of the adenomatous polyposis coli (APC) protein, which reduces the differentiation of intestinal epithelial cells (IECs), leading to the onset of adenoma and CRC progression. Furthermore, the mutations and LOH of *APC* alter the quantitative regulation of the β-catenin protein, which accumulates in the nucleus, favoring the activity of transcription factors for cell proliferation gene expression and reducing differentiation [27]. Familial adenomatous polyposis (FAP) accounts for less than 1% of CRC cases. It is an inherited CRC syndrome caused by a germline mutation in the *APC* gene, inherited in an autosomal dominant pattern. Around 70% of patients with FAP have a family history of colorectal polyps and cancer. FAP is characterized by the growth of many tens to thousands of adenomas in the rectum and colon during the second and third decade of life. APC is essential for IEC homeostasis and its inactivation facilitates tumorigenesis. Indeed, *APC* somatic truncation mutations are observed in more than 90% of human colon cancers [28,29]. *Wnt* ligands may activate the canonical (β-catenin-dependent) and the non-canonical (β-catenin-independent) pathways. They work in concert to maintain the renewal, defense, and metabolic homeostasis of the colon epithelia [30].

Most of the cellular β-catenin is confined to the adherens junctions on the plasma membrane. Cytosolic β-catenin associates in a complex with APC and axis inhibition protein 1 (AXIN1) proteins, which mediate the N-terminal phosphorylation of β-catenin. This event conducts the ubiquitination of β-catenin by the beta-transducin repeat containing E3 ubiquitin protein ligase (β-TRCP) following proteasomal degradation. When Wnt ligands bind to the Frizzled receptors, Dvl/Dsh is phosphorylated and, in turn, recruits AXIN1 and glycogen synthase kinase-3 beta (GSK3β) adjacent to the plasma membrane, thus preventing the building of the degradation complex. Consequently, unphosphorylated β-catenin eludes recognition by β-TRCP and moves into the nucleus. There, it binds to the T-cell factor (TCF) and lymphoid enhancer-binding protein family (LEF) transcription factors.

The activated β-catenin/TCF/LEF complex triggers gene transcription that regulates cell proliferation and survival. In normal cells, two LEF1 isoforms regulate *Wnt*-dependent pathways as apoptosis, motility, and gene transcription, and its expression in human colon tissue gradually increased from a normal colon, low-grade adenoma, high-grade adenoma, to adenocarcinoma [31]. Then, β-catenin accumulates in the cytoplasm and in the nucleus [32].

The B-cell CLL/lymphoma 9 (BCL9) protein is a novel co-factor of canonical Wnt/β-catenin signaling [33–35]. It forms a complex with β-catenin-LEF/TCF to activate the transcription of Wnt target genes, after the hyper-activation of canonical Wnt signaling [36]. In CRC tissues, *Wnt3* is highly expressed to sustain autocrine Wnt activity and CRC progression by EMT and is indicative of advanced stages with poor prognoses [37]. Inhibiting *Wnt3* secretion inhibits colon cancer cells proliferation [38]. The *Wnt3a* expression was also increased and associated with EMT, which is indicative of advanced stages with poor prognoses [39]. Moreover, *Wnt3a* was overexpressed in CRC primary tissues than in metastatic areas, suggesting that *Wnt3a* was expressed early in cancer rather than appearing as it progressed. A more recent study discovered that *Wnt3a* inhibits the ability of human colon myofibroblasts to proliferate and migrate [40]. Thus, various CRC subgroups have distinct molecular and cellular properties contributing to *Wnt3a*'s context-dependent nature.

Wnt5a is a potent non-canonical Wnt ligand that strongly antagonizes and, ultimately, suppresses the functions of canonical Wnt ligands [41]. Recent research shows that the *Wnt5a* non-canonical ligand can act as both a tumor suppressor and an oncogenic agent, promoting and inhibiting tumor processes through canonical and non-canonical Wnt signaling [42,43]. The exact role of *Wnt5a* in CRC is contradictory [44].

As described above, Wnt/β-catenin signaling is well-known in human CRC, but less studied in the adenoma formation in the early stages of colon tumorigenesis. Based on what is reported in the current literature, there are no studies analyzing these aspects on human adenomas. The main aim of this study was to try to fill the gap that exists in the literature on molecular alterations during adenoma formation and in the very early stages of colorectal carcinogenesis. Only the early inactivation of the *APC* gene is known. It is the most studied gene of the Wnt pathway and there are many mutational analysis studies of the *APC* gene but few on gene and protein expression. The first aim of this study was to study the *APC* gene, and then to analyze less studied Wnt pathway genes, such as *Wnt3a*, *Wnt5a*, *LEF1*, and *BCL9*. To complete the study, we analyzed *β-catenin*, whose role in colorectal carcinogenesis is known, but little is known about its role in adenoma formation. Finally, we investigated the relationships among these key elements of Wnt/β-catenin signaling.

The study also aimed to verify whether there were molecular alterations in the tissues surrounding and adjacent to the tumor, to provide clinical insights for disease management.

The analyses of the gene and protein expression were conducted on pathological samples (polyps and related adjacent mucosa) derived from patients both with sporadic adenomas and suffering from FAP.

2. Materials and Methods

2.1. Tissues

Pathological and control tissues were recruited from patients with sporadic adenomas undergoing endoscopic biopsy at the Digestive Endoscopy and Gastroenterology Unit, "Santissima. Annunziata" Hospital of Chieti, while those from patients belonging to families with FAP were recruited at the Surgery Unit, Careggi University Hospital in Florence. The collection of each pathological sample was accompanied by the normal-appearing tissue sample obtained from areas that were at least 5 cm away from the margins of the primary lesion. Finally, normal colonic mucosal samples were collected from healthy donors with age ≥18 years without inflammatory bowel disease and personal or family history of cancer, who had undergone follow-up colonoscopy at the Digestive Endoscopy and Gastroenterology Unit, "Santissima. Annunziata" Hospital of Chieti (UOD of Digestive Endoscopy and Gastroenterology). After the surgical removal, the tissue fragments were stored in RNAlater™ solution (Thermo Fisher Scientific, Waltham, MA, USA) to stabilize the RNA and to preserve proteins, and subsequently stored at −80 °C. The study protocol was approved by the local Ethics Committee and each participant tissue donor provided written informed consent. The study was conducted according to the Declaration of Helsinki and approval was granted from the Institutional Review Board (Prot. Id. RICH1KHE). Adenoma tissues were classified according to conventional histopathological criteria, as defined by World Health Organization (WHO). Patient characteristics and polyp histotype are shown in Table 1.

Table 1. Clinical and histopathological characteristics of 41 patients with FAP and sporadic adenomas analyzed.

			Patients with FAP Polyps			
Case	**Age**	**Sex**	**Phenotype**	**Site and Size of Polyps**	**Dysplasia (L or H)**	***n.* of Polyps**
5FI	25	F	Adenomatous	Diffuse or "carpet", <1 cm	HGD	1060
6FI [a,e]	58	M	Adenomatous	Diffuse	HGD	25
7FI [a,e]	28	F	Adenomatous	Diffuse	HGD	375
8FI [b,e]	18	F	Adenomatous (Tubular–villous)	Diffuse	HGD	415
9FI [c,e]	15	F	Adenomatous	Diffuse	LGD	375
16FI	n.a	F	Adenomatous	Diffuse	LGD	n.a
25FI	n.a.	M	Adenomatous	Diffuse	LGD	n.a.
26FI	46	F	Adenomatous	Diffuse	LGD	835
31FI		M	Adenomatous	Diffuse		
33FI	49	F	Adenomatous	Diffuse	LGD	97
35FI [c,e]	31	M	Adenomatous	Diffuse	LGD	550
36FI	42	F	Adenomatous	Diffuse	LGD	250
39FI	42	M	Adenomatous	Diffuse	LGD	430
40FI [a,d,e]	61	F	Adenomatous	Diffuse	HGD	730
41FI	49	M	Adenomatous	Diffuse	LGD	1025
42FI	42	F	Adenomatous and amartomatous	Diffuse	LGD	210
43FI	36	M	Adenomatous	Diffuse	LGD?	n.a
			Patients with Sporadic Polyps			
Case	**Age**	**Sex**	**Phenotype**	**Site and Size of Polyps**	**Dysplasia (L or H)**	**Morphology**
1CH	50	M	Hyperplastic	Sigma, 6 mm	LGD	Spl
2CH	67	M	Tubular	Sigma, 10 mm	LGD	Spl
3CH	49	M	Hyperplastic	Sigma, 4 mm	LGD	Spl
9CH	47	M	Tubular–villous	Retto, 15 mm	LGD	Ppl
11CH	57	F	Hyperplastic–adenomatous	Descending, 4 mm	LGD	Spl
13CH	83	F	Tubular	Descending, 15 mm	LGD	Spl
15CH	37	F	Villous	Sigma, 50 mm	HGD	Ppl
16CH	60	M	Tubular	Cecum, 15 mm	LGD	Ppl
17CH	66	M	Tubular–villous	Sigma, 15 mm	LGD	Ppl

Table 1. *Cont.*

Patients with Sporadic Polyps						
Case	**Age**	**Sex**	**Phenotype**	**Site and Size of Polyps**	**Dysplasia (L or H)**	**Morphology**
18CH	64	M	Tubular–villous	Descending, 8 mm	LGD	Spl
21CH	78	M	Tubular–villous	Sigma, 10 mm	LGD	Spl
22CH	67	M	Tubular–villous	Rectum, 10 mm	LGD	Spl
23CH	68	F	Tubular–villous	Sigma, 10 mm	LGD	Ppl
24CH	59	M	Tubular–villous	Ascending	LGD	Ppl
25CH	77	M	Tubular–villous	Descending	HGD	Ppl
26CH	69	M	Tubular	Splenic flexure, 10 mm	LGD	Spl
27CH	61	F	Tubular–villous	Sigma, 15 mm	LGD	Ppl
28CH	77	M	Tubular–villous	Hepatic flexure, 5 mm	LGD	Spl
29CH	47	M	Hyperplastic–adenomatous	Descending, 20 mm	Not atypical	Ppl
30CH	53	M	Hyperplastic–adenomatous	Retto-sigma, 7 mm	Not atypical	Spl
31CH	76	M	Tubular	Ascending, 5 mm	LGD	Spl
32CH	51	M	Tubular–villous	Ascending, 45 mm	LGD	Spl
33CH	68	F	Tubular–villous	Colon, 40 mm	LGD	LST-G
34CH	67	M	Tubular	Colon sx, 7 mm	LGD	Ppl

[a] Presence of rectal cancer; [b] APC mutation: c.4666_4665ins(p.Thr1556fs); [c] APC mutation: c.2805 C>(p.Tyr935X); [d] APCmutation: c.3927_3931del(p.Glu1309_Asp.fsx1312); [e] [45]. Ppl: pedunculated polypoid lesion; Spl: sessile polypoid lesion; LST-G: laterally spreading tumor (LST)—granular shape (G); HGD: high-grade dysplasia; LGD: low-grade dysplasia, FAP: familial adenomatous polyposis; FI: case from hospital of Florence; CH: case from hospital of Chieti.

Tissues were selected based on RNA availability. Samples with insufficient quantity of target in the cDNA template or protein were not included in the gene and protein expression analyses. The study included 68 colonic samples; 58 biopsies (33 polyps and 25 adjacent mucosa) belonged to 41 patients, of which 17 were affected by FAP, and ten normal colonic mucosal samples were collected from 10 healthy donors.

2.2. Real-Time Quantitative PCR Analysis (qRT-PCR)

Total RNA was separated from colon tissues homogenized in liquid nitrogen with a mortar and pestle, using TRIzol® Reagent (Applied Biosystems, Thermo Fisher Scientific, Waltham, MA, USA) following the manufacturer's instructions at RNase-free atmosphere. The RNA samples were assessed for purity and quantified using a Nanodrop 1000 Spectrophotometer (Applied Biosystems, Thermo Fisher Scientific, Waltham, MA, USA). Complementary DNA (cDNA) were synthesized as previously described [46]. The mRNA levels were evaluated by SYBR Green and TaqMan assay by quantitative real-time PCR (qRT-PCR) analysis using StepOne™ 2.0 (Applied Biosystems, Thermo Fisher Scientific, Waltham, MA, USA). Data were analyzed by the comparative Ct method and graphically represented as $2^{-\Delta\Delta Ct} \pm$ SD. In accordance with this method, the mRNA amounts of the target genes were normalized by the ratio on the median value of the endogenous housekeeping gene (GUSB). Primer sequences are available in Catalano et al., 2021 [46]. The cycling conditions were performed as follows: 10 min at 95 °C and 40 cycles of 15 s at 95 °C, followed by 1 min at 60 °C, and final elongation of 15 s at 95 °C. All data were validated in a second analysis. The sequences of paired oligonucleotides were as follows:

- 5′-GCTTGATAGCTACAAATGAGGACC-3′ and 5′-CCACAAAGTTCCACATGC-3′ for APC; RefSeq: [NM_000038];
- 5′-CATGAACCGCCACAACAAC-3′ and 5′-TGGCACTTGCACTTGAGGT-3′ for *WNT-3a*; RefSeq: [NM_033131];
- 5′-CTCATGAACCTGCACAACAACG-3′ and 5′-CCAGCATGTCTTCAGGCTACAT-3′ for *WNT-5a*; RefSeq: [NM_03392];
- 5′-CCAACTTGCCATCAATGAATAA-3′ and 5′-GGCATCTGATTGGAGTGAGAA-3′ for BCL-9; RefSeq: [NM_004326];
- 5′-GAC GAG ATG ATC CCC TTC AA-3′ and 5′-AGG GCT CCT GAG AGG TTT GT-3′ for LEF-1; RefSeq: [NM_016269];
- 5′-AGCCAGTTCCTCATCAATGG-3′ and 5′-GGTAGTGGCTGGTACGGAAA-3′ for GUSB; RefSeq: [NM_000181].

2.3. Western Blotting

Total proteins were isolated from pathological and control tissues homogenized in liquid nitrogen with a mortar and pestle, using RIPA lysis buffer (Cell Signaling Technology, Beverly, MA, USA). Proteins were quantified by Bradford Assay (Thermo Fisher Scientific, Waltham, CA, USA) and the protein lysates were subjected to electrophoresis, followed by immunoblotting. Then, 30 µg of total proteins was incubated with SDS-PAGE sample buffer (125 mM Tris-HCl, pH 6.8, 4% SDS, 20% glycerol, 10% beta-mercaptoethanol, and 0.004% bromophenol blue) at 100 °C for 5 min. Tris-glycine SDS running buffer (25 mM Tris, 250 mM glycine, 0.1% SDS) was used for electrophoresis. Proteins were separated for 60′ at 100 V in an 8% Tris/glyicine acrylamide gel (Mini PROTEAN electrophoresis cell). After electrophoresis, proteins were transferred onto the Immobilon-P PVDF membrane at 80 mA for 1 h by using Tris-glycine SDS (48 mM Tris, 39 mM glycine, 0.037% SDS) transfer buffer with 20% methanol in a Mini Trans-Blot Electrophoretic Transfer Cell. The blotted PVDF membranes were directly blocked with 5% bovine serum albumin (BSA) in Tris-buffered saline with 0.05% Tween-20 (TBST) for 1 h and incubated overnight 4 °C with primary antibodies diluted in TBST with 5% BSA. Primary antibodies were Active-β-Catenin (1:1000) (Cell Signaling Technology), APC (1:1000) (Merck-Millipore, Burlington, MA, USA), and β-actin (1:8000) (Sigma-Aldrich, St. Louis, MI, USA), used as a protein loading control. The blotted membranes were washed thoroughly with TBST before incubation with diluted (1:10,000) HRP-conjugated anti-rabbit or anti-mouse (Bethyl Laboratories, Montgomery, TX, USA). The immune complexes were visualized using the ECL western blot detection system (EuroClone) by using AllianceLD2 hardware (UVItec Limited, Cambridge, UK).

2.4. Statistical Analysis

All measurements were made after three independent experiments were conducted under the same experimental condition. A mean value of all experiments plus standard deviation *(SD)* was shown for qRT-PCR data, while, for western blotting, a representative value was shown. Statistical analyses were performed using SPSS software 29.0. Kruskal–Wallis non-parametric test for independent samples was used to compare the expression of the five genes in polyps, adjacent mucosa, and normal tissue for each gene separately. Spearman's ρ correlation coefficients were evaluated between the five genes. p-value was considered as significant if <0.05. Microsoft Excel version 16.66.1 was used to draw a scatter plot and linear trend line. Descriptive statistics were carried out using GraphPad Prism version 9.0 (GraphPad Software Inc., La Jolla, CA, USA).

3. Results

We analyzed the expression of five genes of the Wnt/β-catenin family in all available tissues, grouping them into three categories: colorectal adenomas, adjacent mucosa, and normal tissues. To evaluate any alterations in the protein expression of APC and β-catenin in the mucosa adjacent to the adenoma transition, we performed protein expression in the cases in which the matched adenoma and adjacent mucosal tissues were available.

3.1. Gene Expression of APC, Wnt3a, Wnt5a, BCL9, and LEF1 in FAP and Tubular–Villous Adenomas

The gene expression of the five Wnt/β-catenin signaling genes (*APC*, *BCL9*, *LEF1*, *Wnt3a*, and *Wnt5a*) was conducted on 52 colonic samples; 42 biopsies (25 polyps and 17 adjacent mucosa) belonged to 33 patients, of which 15 were affected by FAP, and 10 were normal colonic mucosal samples from 10 healthy donors.

The expressions of *APC*, *Wnt3a*, *Wnt5a*, *BCL9*, and *LEF1* in the polyps, adjacent mucosa, and normal tissue were compared by the Kruskal–Wallis non-parametric test for independent samples. Significant results were found for *APC* ($p = 0.03$) and *Wnt5a* ($p = 0.01$). A post hoc test showed a significant difference in *APC* expression for polyps compared to healthy mucosa ($p = 0.03$) (Figure 1), and in *Wnt5a* expression for healthy mucosa compared to both polyps ($p = 0.04$) and adjacent mucosa ($p = 0.01$).

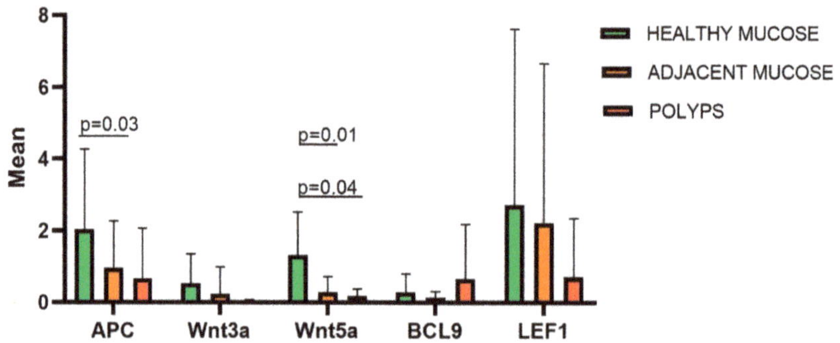

Figure 1. The figure shows the mean expression for five Wnt/β-catenin pathway genes (*APC*, *Wnt3a*, *Wnt5a*, *BCL9*, and *LEF1*) in healthy colorectal mucosa (n.10), adjacent mucosa (n.17), and polyps (n.25). Significant differences were detected in *APC* and *Wnt5a* expressions in polyps compared to normal tissue ($p = 0.03$), and, for *Wnt5a* expression, also in polyps compared to normal tissue ($p = 0.04$). *APC*: adenomatous polyposis coli; *Wnt3a*: *Wnt* family member 3a; *Wnt5a*: *Wnt* family member 5a; *BCL9*: B-cell CLL/lymphoma 9; *LEF1*: lymphoid enhancer-binding factor 1.

Furthermore, the Spearman's ρ analysis revealed significant correlations between *Wnt3a* and *BCL9* ($\rho = 0.34$, $p = 0.03$) when the polyps and adjacent mucosa were considered in the same category, and between *Wnt5a* and *LEF1* ($\rho = 0.70$, $p = 0.03$) for healthy tissues. A scatter plot and linear trend line are shown in Figure 2.

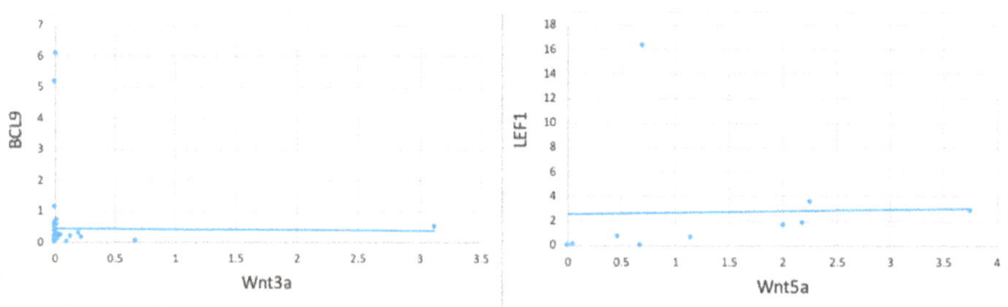

(a) (b)

Figure 2. Correlation between *Wnt3a* vs. *BCL9* and *Wnt5a* vs. *LEF1* gene expression in 52 colon tissue samples. Correlation is significant at the 0.05 level (2-tailed). The figure represents a scatter plot and linear trend line of the expression values between *Wnt3a* vs. *BCL9* in polyps and adjacent mucosa (n. 42) (panel (**a**)) and of *Wnt5a* vs. *LEF1* in normal colonic mucosa (n. 10) (panel (**b**)).

The gene expression results were graphically represented using a heat map. *APC, Wnt3a, Wnt5a, BCL9,* and *LEF1* tend to reduce from healthy to pathological tissue (Figure 3). *APC, Wnt5a,* and *LEF1* are the most expressed genes in healthy mucosa (Figure 3).

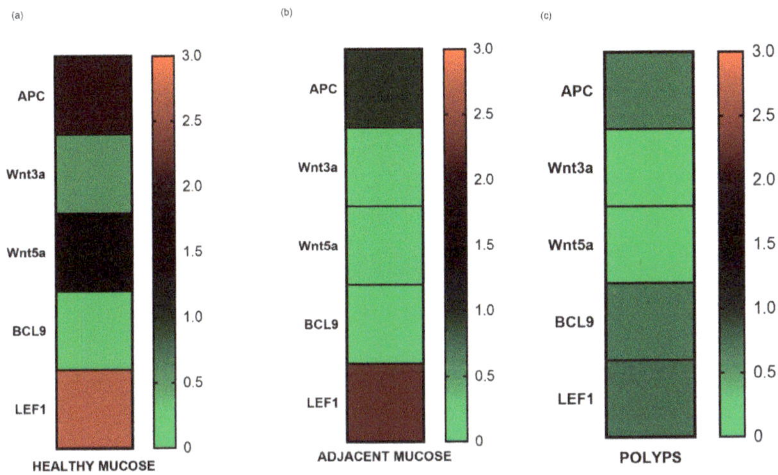

Figure 3. The figure represents a heat map of the gene expression means of *APC, Wnt3a, Wnt5a, BCL9,* and *LEF1* in healthy mucosa (**a**), adjacent mucosa (**b**), and polyps (**c**). It is shown how gene expression tends to reduce from healthy to pathological tissue.

3.2. Protein Expression Assay by Western Blotting of APC and β-catenin

A 50% reduction of APC gene expression in the adjacent mucosa compared to the healthy mucosa, and a more accentuated reduction in polyps, have been detected (Figure 1). Therefore, we wanted to investigate the levels of APC and β-catenin proteins in still available tissues. Then, a western blot analysis of matched polyp-adjacent mucosal tissues was performed. The matched adenoma and adjacent mucosal tissues were available for six FAP patients and for seven patients with sporadic adenomas. The analyses revealed the expression of the full-length APC protein (300 kDa band) in all matched adenoma-adjacent mucosa samples (Figures 4a and 5a). A decrease in APC expression is visible in all the adenomas, both familial and sporadic, compared to adjacent mucosa. As we expected, as regards the β-catenin expression, its active form appeared to be increased in all the familial and sporadic adenomas compared to adjacent mucosa (Figures 4a and 5a). Sample 42FI-P does not show a decrease in APC expression, and, intriguingly, it shows a reduced active β-catenin expression, also visible in the matched adjacent mucosa. Moreover, for sample 43FI, a reduced expression of the active β-catenin was noted, apparently not associated with APC modulation. The relative expression quantification of the selected proteins (APC and active β-catenin) detected by WB in the available matched adenoma and adjacent mucosal tissues are shown in Figures 4b,c and 5b,c.

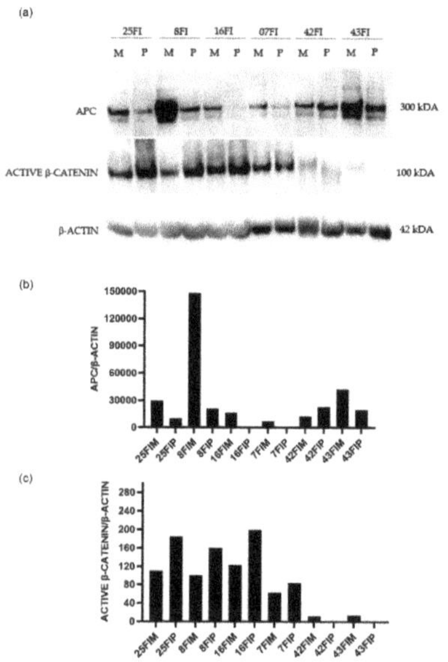

Figure 4. Western blotting analysis in familial adenomas determining the protein expression levels of APC and β-catenin in polyp (P) vs. adjacent mucosa (M). Data shown are representative of three independent experiments. The expression levels of panel (**a**) were determined by densitometric analysis (panel (**b**,**c**)) and calculated in relation to the β-actin level. kD: kilodalton as protein molecular weight unit.

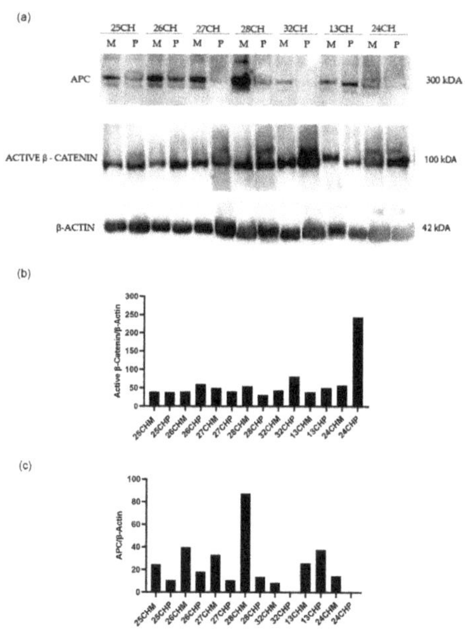

Figure 5. Western blotting analysis in sporadic adenomas determining the protein expression levels of APC and β-catenin in polyp (P) vs. adjacent mucosa (M). Data shown are representative of three independent experiments. The expression levels of panel (**a**) were determined by densitometric analysis (panel (**b**,**c**)) and calculated in relation to the β-actin level. kD: kilodalton as protein molecular weight unit.

4. Discussion

The current literature lacks significant Wnt/β-catenin/APC gene expression studies on adenomas. Our study is the first to analyze the modulation of the Wnt/β-catenin pathway in early carcinogenesis. Understanding the significance of benign pre-cancerous lesions is widely recognized; yet, studying them is complicated in terms of sampling, mainly due to the difficulty in obtaining biopsies and their small size. The long period required for polyps and cancer to develop, as well as the tendency of early-stage cancers to be asymptomatic in many individuals, allows a window of opportunity for polypectomy and cancer prevention, as well as for early diagnosis and highly effective drug administration for early-stage cancers. Blood-based tests could detect genetic changes associated with polyps released into circulation. Promoting the uptake and completion of follow-up testing and treatment holds significant potential to save lives. Therefore, understanding the progression from colorectal adenomas to colorectal carcinomas is essential for the development of new therapeutic strategies and improving clinical outcomes.

Tumor and adjacent mucosa biopsies are the definitive diagnostic procedures for a histological evaluation. The adjacent mucosa, also known as "apparently healthy" mucosa, being closely associated with polyps in the tumor microenvironment (TME), may have already been impacted molecularly [47]. Our data support this hypothesis.

Initially, the study was designed to analyze any molecular differences between sporadic adenomas and FAP, but the sample size did not allow this. It is already known that *APC* is inactivated in FAP polyps. This study focused on evaluating the APC gene and protein expression not only on pathological FAP tissue and tubular–villous adenomas, but also on apparently normal endoscopic mucosal tissue sampled at the 5 cm proximal. Very interestingly, a lower APC expression was detected in the adjacent mucosa compared to the normal one, but, at the same time, it is the most expressed gene in the healthy mucosa together with *LEF1* and *Wnt5a*. This is explained because it is the regulator of the Wnt pathway in a colonic tissue adjacent to the tumor but in which it still retains its functions.

APC has a functional role in the canonical (β-catenin-dependent) Wnt signaling pathways [48]. Most cases of CRCs are initiated by *APC*-inactivating mutations, leading to the constitutive activation of the Wnt/β-catenin signaling pathway. Most of the literature has extensively assessed the role of APC somatic and germline mutations in familial as well as sporadic forms of CRC.

It is known that *APC* is inactivated in FAP polyps. Our results showed a reduction in APC expression both in FAP and sporadic polyps. Additionally, a lower APC expression was detected in the adjacent mucosa.

Moreover, APC can inhibit the initiation and development of colorectal tumor, independently of canonical Wnt signaling. APC assists in chromosome segregation, establishes cellular polarity and migration, and represses DNA replication [27]. *APC* mutations contribute in early adenoma creation leading to chromosomal instability by triggering spindle abnormalities and the deregulation of microtubules/the actin cytoskeleton. Moreover, *APC* mutations increase cell migration by reducing cell adhesion via the deregulation of β-catenin and E-cadherin distributions among the cytoplasm and the cell membrane [49]. Xu et al. demonstrated that the overexpression of the erythropoietin-producing hepatocyte (EphB6), a member of the tyrosine kinase family, along with *APC* gene mutations, increases proliferation, migration, and invasion in the colon epithelial cell line, IMCE, supporting the role of *APC* mutations in promoting tumorigenesis in CRC [50]. In our study, the APC gene was less expressed in the colon tumor compared to the adjacent colonic mucosa and to the mucosa of healthy controls. Moreover, the APC protein was less expressed in the colon tumor compared to the adjacent colonic mucosa. All results confirmed the literature data on carcinomas.

LEF1 is a downstream mediator and nuclear effector of the Wnt/β-catenin signaling pathway [51]. In normal cells, two LEF1 isoforms are in regulation of *Wnt*-dependent pathways as apoptosis, motility, and gene transcription. The role of LEF1 is controversial, usually detected and upregulated in most colonic carcinomas, enhancing the progression [52]. Nevertheless, little is known about the expression of LEF1 in early carcinogenesis. LEF1 is frequently highly expressed in tumor development, potentially driving cancer proliferation and spread [53]. Reducing LEF1 in glioblastoma multiforme cells restricts their ability to invade, migrate, and proliferate, as well as the self-renewal capability of stem-like cells [54]. Myc controls the expression of LEF1 to activate the Wnt pathway in colon cancer [55]. The enzyme lysine-specific demethylase 1 (LSD1) also promotes bladder cancer progression by enriching LEF1 expression and improving EMT [56]. Although the LEF1 expression may play a role in cancer development [57–59], insufficient evidence supports its involvement in malignant phenotype changes. The molecular mechanisms regulating LEF1 in the advancement of colonic adenocarcinoma are currently unknown. However, a study by Xiao et al. in 2021 [52] keeps LEF1 as a potential therapeutic target for colonic adenocarcinoma, suggesting that it enhances the motility of cancer cells by reshaping the lamellipodia/filopodia and the polymerization of F-actin/β-tubulin. The Xiao et al., 2021 study findings support LEF1 as a potentiator and potential therapeutic target for colonic adenocarcinoma. LEF1 increases colonic adenocarcinoma cells' motility by remodeling the lamellipodia/filopodia and the polymerization of F-actin/β-tubulin. LEF1 maintains the viability and growth of colonic adenocarcinoma cells through improving proliferation and Lamin B1 expression, and reducing apoptosis. Furthermore, LEF1 is nearly linked with EMT. However, an in vivo study published in Science [60] showed that *LEF1* restricts ectopic crypt formation and tumor cell growth in colon adenomas from APC-deficient mice. The loss of *Lef1* markedly increased tumor initiation and tumor cell proliferation, reduced the expression of several Wnt antagonists, and increased the *Myc* proto-oncogene expression and the formation of ectopic crypts in *Apc*-mutant adenomas. These results uncover a previously unknown negative feedback mechanism in CRC, in which the ectopic *Lef1* expression suppresses intestinal tumorigenesis by restricting adenoma cell dedifferentiation to a crypt-progenitor phenotype and by reducing the formation of cancer stem cell niches. The recent literature data therefore demonstrate how the controversy on the function of *LEF1* in colorectal tumorigenesis is still open.

There are many articles in the literature that underline the role of *LEF1* as a transcription factor and, therefore, its increased presence in cancer. But there are others, such as the one published in Science [60] in which its loss increased cell growth. In our study, *LEF1* is highly expressed in the adjacent mucosa and decreased in the polyp, supporting the recent hypothesis of LEF1's role as a tumor suppressor, already in adenomas, but also its essential role in stem cell maintenance during crypt formation and organ development.

BCL9 is considered a key developmental regulator and a well-established oncogenic driver in multiple cancer types, mainly through potentiating the Wnt/β-catenin signaling. BCL9 is recognized as a crucial part of the nuclear β-catenin complexes [61]. It serves as an adapter protein that provides binding sites for the Wnt signaling transcriptional system [62]. BCL9 functions as an oncoprotein by supporting cancer progression primarily through maintaining cancer cell division [63], promoting proliferation and migration, inhibiting apoptosis [64], remodeling the tumor microenvironment and immune system, and regulating the chromosomal instability and karyotype for tumor evolution [65].

Wnt3a is a ligand that triggers the canonical Wnt pathway. In glioblastoma [66], breast and prostate tumors [67,68], and malignant mesothelioma [69], *Wnt3a* stimulates the development of cancer. Furthermore, research has indicated that *Wnt3a* is a tumor suppressor [70]. Marit et al. noted that *Wnt3a* suppresses the growth of multiple B-acute lymphoblastic leukemia cell lines [71]. Qi L. et al. [39] examined the expression of *Wnt3a* in numerous colon cancer tissue samples to investigate its impact on colon cancer progression. They found a strong connection between the *Wnt3a* expression and histological differentiation, clinical stages, metastasis, and recurrence. The data obtained indicate that the upstream component of the Wnt signaling pathway could have a significant impact on the advancement of colon cancer, which is consistent with a recent investigation on colorectal cancer that found high levels of *Wnt3a* expression in both primary and metastatic locations, and its strong correlation with the presence of the metastasis-related protein matrix metalloproteinase (MMP)-9 [72]. In our study, the correlation between BCL9 and *Wnt3a* both in polyps and in the adjacent mucosa confirms two aspects: the first is that the canonical Wnt pathway is already activated in the adjacent mucosa; the second is that it is also activated in early carcinogenesis. This correlation concerns the adenoma formation, regardless of the clinical characteristics of the patient, FAP or sporadic. Finally, our results confirm *Wnt3a* as a tumor suppressor.

The precise role of *Wnt5a* still needs to be supported by conclusive evidence. Specific research indicates that *Wnt5a* is a tumor suppressor, while others propose the contrary. In the progression of colorectal cancer, *Wnt5a* demonstrates diverse roles across various signal transduction pathways [73]. The mRNA of *Wnt5a* is present in most healthy tissues, including the colon. However, its expression significantly increases during the transition from normal tissue to carcinomas [74]. On the other hand, the presence of the *Wnt5a* protein appears to decrease, as it is commonly deactivated in colorectal cancer (CRC) through tumor-specific methylation. This makes it a potential biomarker [75]. *Wnt5a* is believed to function as a tumor suppressor in CRC by inhibiting the Wnt/β-catenin signaling pathway [75].

Traditionally, *Wnt5a* has been considered to be the non-canonical Wnt ligand and triggers Ca2+-dependent effectors and other non-canonical pathways through small Rho-GTPases and c-Jun-NH2-kinase [76]. However, its role in the progression of CRC is intricate and appears to be paradoxical. Multiple studies have demonstrated that *Wnt5a* is silenced in the majority of CRC cell lines and samples because of the frequent methylation in its promoter region [77,78]. The expression of *Wnt5a* showed a negative association with the level of tumor differentiation and the aggressive behavior [77,79].

Meanwhile, the methylation of the *Wnt5a* promoter was closely linked to the microsatellite instability status of colorectal cancer (CRC) patients, and various histone modifications of *Wnt5a* participated in its suppression and could potentially encourage the spread of colon cancer, suggesting that epigenetic processes might improve the *Wnt5a*-mediated signaling in CRC [80,81]. Conversely, other research indicated that *Wnt5a* was consistently overexpressed in intestinal polyps and tumor samples, and elevated *Wnt5a* levels were associated with early recurrence or metastasis in colon cancer patients [82,83]. *Wnt5a* was also found to facilitate the movement of CRC cells by activating Fzd7-driven non-canonical Wnt signaling and enhance CRC cells' stemness by activating canonical Wnt signaling [84,85].

Our study showed *Wnt5a* expressed in healthy mucosa. The correlation between *Wnt5a* and LEF1 detected in healthy tissues could indicate that the Wnt non-canonical pathway is active in normal colonic tissue. Furthermore, these results could also denote a possible inflammatory state of donors undergoing a screening colonoscopy, as both *Wnt5a* and LEF-1 are linked to the inflammatory state. Aberrant Wnt signaling is linked to defects in the chronic inflammatory response. Indeed, in the still normal colonic mucosa, various inflammatory mediators can actively contribute to the creation of a TME favorable to cell transformation, survival, and proliferation [86]. The mutual interaction of epithelial cells within the TME influences the stages of tumorigenesis driven, to a large extent, by inflammation. This may be an attractive therapeutic target to control inflammation in the colonic mucosa [87]. Furthermore, IEC can drive the plasticity of stromal cells in the TME under the influence of extrinsic factors, such as diet, and the microbiota composition contributing to inflammation and tumorigenesis in CRC [86,88]. However, understanding the interactions between Wnt signaling and inflammatory/immune responses in tumor onset remains a necessary goal for both prevention and therapy, given that the majority of CRCs appear to be immunologically "cold" tumors unresponsive to therapies with immune checkpoint inhibitors [89].

The observed correlations between *Wnt3a* and *BCL9* gene expressions and between *Wnt5a* and *LEF1* could provide biological significance such as the activation of the canonical *Wnt* pathway in pathological tissues and the non-canonical pathway in healthy tissues, although they require further studies.

APC and β-catenin were downregulated and upregulated, respectively, already in the adenoma formation. As regards *Wnt3a, Wnt5a, BCL9,* and *LEF*, the study underlined a modulation of these genes in pathological tissue compared to normal tissue.

The study showed that the mucosa adjacent to adenoma is already altered at a molecular level. This result provides a new clinical insight: verifying the nature of the colorectal mucosa during follow-up by analyzing markers such as APC and β-catenin.

These findings provide information on the possible progression towards carcinoma of the residual mucosa, already altered at a molecular level. The investigation would take place during follow-up and, therefore, would not add further medicalization to the patient. To contain costs, it could initially be aimed at FAP patients, who have a greater risk of recurrence of adenomas, rather than sporadic patients.

Limitations

This is an observational study, and additional investigations are needed to facilitate the understanding of the potential clinical implications of the results. *APC* has been validated with two independent techniques, qRT-PCR and WB. It has not been possible to validate the results from the analyses of *Wnt3a, Wnt5a, BCL9,* and *LEF1* with another independent technique, due to the small size of the biopsies, which, therefore, constituted a limiting factor. The results regarding these four genes should be confirmed on a larger sample, before hypothesizing about their use as markers in the clinic. There is very little scientific literature concerning both carcinomas and adenomas, and even validation using datasets such as The Human Protein Atlas (https://www.proteinatlas.org, accessed on 29 July 2024) has not provided us with the possibility of a comparison.

5. Conclusions

This is the first study analyzing the gene expression of *APC, Wnt3A, Wnt5A, BCL9,* and *LEF1* in the colon polyps vs. adjacent mucosa and vs. normal mucosa from control individuals. These findings of altered expression levels of Wnt genes in apparently normal adjacent mucosa from patients with familial and sporadic colon polyps underline an interplay between the tumor and the surrounding colonic epithelium. This may aid in identifying patients at risk of developing cancer.

In conclusion, our study aims to enhance the comprehension of the pathogenesis in colorectal adenomas (CRAs) and to propose the utilization of APC and β-catenin as markers in clinical settings. Identifying crucial genes, investigating their potential role in the pathogenesis of colorectal adenomas, and developing gene-targeted medications are pressing clinical and scientific issues that need to be addressed.

Author Contributions: Conceptualization, M.C.C. and G.M.A.; methodology, D.L.D., F.F. (Fabiana Fantini), C.M. and A.F.; recruitment of patients: R.V., F.F. (Ferdinando Ficari), S.S., K.E. and M.N.; writing—review and editing, M.C.C., D.L.D. and G.M.A.; supervision, M.C.C. All authors have read and agreed to the published version of the manuscript.

Funding: This research received no external funding.

Institutional Review Board Statement: The study was conducted in accordance with the Declaration of Helsinki, and approved by the Ethics Committee of the provinces of Chieti and Pescara (protocol code rich1k2he and approved on 27 November 2014).

Informed Consent Statement: Informed consent was obtained from all subjects involved in the study.

Data Availability Statement: The original contributions presented in the study are included in the article; further inquiries can be directed to the corresponding author.

Conflicts of Interest: The authors declare no conflicts of interest.

References

1. Bonnington, S.N.; Rutter, M.D. Surveillance of colonic polyps: Are we getting it right? *World J. Gastroenterol.* **2016**, *22*, 1925–1934. [CrossRef] [PubMed] [PubMed Central]
2. Click, B.; Pinsky, P.F.; Hickey, T.; Doroudi, M.; Schoen, R.E. Association of Colonoscopy Adenoma Findings with Long-term Colorectal Cancer Incidence. *JAMA* **2018**, *319*, 2021–2031. [CrossRef] [PubMed]
3. Eide, T.J. Risk of colorectal cancer in adenoma-bearing individuals within a defined population. *Int. J. Cancer* **1986**, *38*, 173–176. [CrossRef] [PubMed]
4. Fearon, E.R.; Vogelstein, B. A genetic model for colorectal tumorigenesis. *Cell* **1990**, *61*, 759–767. [CrossRef] [PubMed]
5. Yang, B.; Mao, L.; Li, Y.; Li, Q.; Li, X.; Zhang, Y.; Zhai, Z. β-catenin, leucine-rich repeat-containing G protein-coupled receptor 5 and GATA-binding factor 6 are associated with the normal mucosa-adenoma-adenocarcinoma sequence of colorectal tumorigenesis. *Oncol. Lett.* **2018**, *15*, 2287–2295. [CrossRef]
6. Kim, N.H.; Jung, Y.S.; Jeong, W.S.; Yang, H.-J.; Park, S.-K.; Choi, K.; Park, D.I. Miss rate of colorectal neoplastic polyps and risk factors for missed polyps in consecutive colonoscopies. *Intest. Res.* **2017**, *15*, 411–418. [CrossRef]
7. International Agency for Research on Cancer. Colorectal cancer screening. In *IARC Handbooks of Cancer Prevention*; International Agency for Research on Cancer: Lyon, France, 2019; Volume 17, p. 300.
8. Lieberman, D.A.; Rex, D.K.; Winawer, S.J.; Giardiello, F.M.; Johnson, D.A.; Levin, T.R. Guidelines for colonoscopy surveillance after screening and polypectomy: A consensus update by the US Multi-Society Task Force on Colorectal Cancer. *Gastroenterology* **2012**, *143*, 844–857. [CrossRef]
9. Aceto, G.M.; Catalano, T.; Curia, M.C. Molecular Aspects of Colorectal Adenomas: The Interplay among Microenvironment, Oxidative Stress, and Predisposition. *BioMed Res. Int.* **2020**, *2020*, 1726309. [CrossRef] [PubMed] [PubMed Central]
10. Dornblaser, D.; Young, S.; Shaukat, A. Colon polyps: Updates in classification and management. *Curr. Opin. Gastroenterol.* **2023**, *40*, 14–20. [CrossRef] [PubMed]
11. Galuppini, F.; Fassan, M.; Mastracci, L.; Gafà, R.; Mele, M.L.; Lazzi, S.; Remo, A.; Parente, P.; D'amuri, A.; Mescoli, C.; et al. The histomorphological and molecular landscape of colorectal adenomas and serrated lesions. *Pathologica* **2021**, *113*, 218–229. [CrossRef] [PubMed] [PubMed Central]
12. Dubé, C.; Yakubu, M.; McCurdy, B.R.; Lischka, A.; Koné, A.; Walker, M.J.; Peirson, L.; Tinmouth, J. Risk of advanced adenoma, colorectal cancer, and colorectal cancer mortality in people with low-risk adenomas at baseline colonoscopy: A systematic review and meta-analysis. *Am. J. Gastroenterol.* **2017**, *112*, 1790–1801. [CrossRef] [PubMed]
13. Lucci-Cordisco, E.; Risio, M.; Venesio, T.; Genuardi, M. The growing complexity of the intestinal polyposis syndromes. *Am. J. Med. Genet. Part A* **2013**, *161*, 2777–2787. [CrossRef] [PubMed]
14. Corley, D.A.; Jensen, C.D.; Marks, A.R.; Zhao, W.K.; de Boer, J.; Levin, T.R.; Doubeni, C.; Fireman, B.H.; Quesenberry, C.P. Variation of adenoma prevalence by age, sex, race, and colon location in a large population: Implications for screening and quality programs. *Clin. Gastroenterol. Hepatol.* **2013**, *11*, 172–180. [CrossRef]
15. Pommergaard, H.-C.; Burcharth, J.; Rosenberg, J.; Raskov, H. The association between location, age and advanced colorectal adenoma characteristics: A propensity-matched analysis. *Scand. J. Gastroenterol.* **2016**, *52*, 1218929. [CrossRef]
16. Klein, J.L.; Okcu, M.; Preisegger, K.H.; Hammer, H.F. Distribution, size and shape of colorectal adenomas as determined by a colonoscopist with a high lesion detection rate: Influence of age, sex and colonoscopy indication. *United Eur. Gastroenterol. J.* **2016**, *4*, 438–448. [CrossRef]
17. Ijspeert, J.E.; van Doorn, S.C.; van der Brug, Y.M.; Bastiaansen, B.A.; Fockens, P.; Dekker, E. The proximal serrated polyp detection rate is an easy-to-measure proxy for the detection rate of clinically relevant serrated polyps. *Gastrointest. Endosc.* **2015**, *82*, 870–877. [CrossRef]
18. Kim, K.-H.; Kim, K.-O.; Jung, Y.; Lee, J.; Kim, S.-W.; Kim, J.-H.; Kim, T.-J.; Cho, Y.-S.; Joo, Y.-E. Clinical and endoscopic characteristics of sessile serrated adenomas/polyps with dysplasia/adenocarcinoma in a Korean population: A Korean Association for the Study of Intestinal Diseases (KASID) multicenter study. *Sci. Rep.* **2019**, *9*, 3946. [CrossRef]
19. Kahi, C.J.; Hewett, D.G.; Norton, D.L.; Eckert, G.J.; Rex, D.K. Prevalence and variable detection of proximal colon serrated polyps during screening colonoscopy. *Clin. Gastroenterol. Hepatol.* **2011**, *9*, 42–46. [CrossRef] [PubMed]
20. Leggett, B.; Whitehall, V. Role of serrated pathway in colorectal cancer pathogenesis. *Gastroenterology* **2010**, *138*, 2088–2100. [CrossRef]
21. Lochhead, P.; Chan, A.T.; Giovannucci, E.; Fuchs, C.S.; Wu, K.; Nishihara, R.; O'Brien, M.; Ogino, S. Progress and opportunities in molecular pathological epidemiology of colorectal premalignant lesions. *Am. J. Gastroenterol.* **2014**, *109*, 1205–1214. [CrossRef] [PubMed]
22. Dinarvand, P.; Davaro, E.P.; Doan, J.V.; Ising, M.E.; Evans, N.R.; Phillips, N.J.; Lai, J.; Guzman, M.A. Familial adenomatous polyposis syndrome: An update and review of extraintestinal manifestations. *Arch. Pathol. Lab. Med.* **2019**, *143*, 1382–1398. [CrossRef] [PubMed]
23. Taherian, M.; Lotfollahzadeh, S.; Daneshpajouhnejad, P.; Arora, K. Tubular Adenoma. 3 June 2023. In *StatPearls [Internet]*; StatPearls Publishing: Treasure Island, FL, USA, 2024. [PubMed]
24. Tischoff, I.; Tannapfel, A. Präkanzerosen im Kolon. *Der Internist* **2013**, *54*, 691–698. [CrossRef] [PubMed]
25. Zhang, L.; Shay, J.W. Multiple Roles of APC and its Therapeutic Implications in Colorectal Cancer. *J. Natl. Cancer Inst.* **2017**, *109*, djw332. [CrossRef] [PubMed] [PubMed Central]

26. Schatoff, E.M.; Leach, B.I.; Dow, L.E. WNT Signaling and Colorectal Cancer. *Curr. Color. Cancer Rep.* **2017**, *13*, 101–110. [CrossRef] [PubMed] [PubMed Central]
27. Hankey, W.; Frankel, W.L.; Groden, J. Functions of the APC tumor suppressor protein dependent and independent of canonical WNT signaling: Implications for therapeutic targeting. *Cancer Metastasis Rev.* **2018**, *37*, 159–172. [CrossRef] [PubMed] [PubMed Central]
28. Fodde, R.; Smits, R.; Clevers, H. APC, Signal transduction and genetic instability in colorectal cancer. *Nat. Rev. Cancer* **2001**, *1*, 55–67. [CrossRef] [PubMed]
29. Morin, P.J.; Sparks, A.B.; Korinek, V.; Barker, N.; Clevers, H.; Vogelstein, B.; Kinzler, K.W. Activation of beta-catenin-Tcf signaling in colon cancer by mutations in beta -catenin or APC. *Science* **1997**, *275*, 1787–1790. [CrossRef] [PubMed]
30. Steinhart, Z.; Angers, S. Wnt signaling in development and tissue homeostasis. *Development* **2018**, *145*, dev146589. [CrossRef] [PubMed]
31. Han, W.; He, L.; Cao, B.; Zhao, X.; Zhang, K.; Li, Y.; Beck, P.; Zhou, Z.; Tian, Y.; Cheng, S.; et al. Differential expression of LEF1/TCFs family members in colonic carcinogenesis. *Mol. Carcinog.* **2017**, *56*, 2372–2381. [CrossRef]
32. Brembeck, F.H.; Wiese, M.; Zatula, N.; Grigoryan, T.; Dai, Y.; Fritzmann, J.; Birchmeier, W. BCL9-2 Promotes Early Stages of Intestinal Tumor Progression. *Gastroenterology* **2011**, *141*, 1359–1370.e3. [CrossRef]
33. Brembeck, F.H.; Schwarz-Romond, T.; Bakkers, J.; Wilhelm, S.; Hammerschmidt, M.; Birchmeier, W. Essential role of BCL9-2 in the switch between β-catenin's adhesive and transcriptional functions. *Genes Dev.* **2004**, *18*, 2225–2230. [CrossRef] [PubMed] [PubMed Central]
34. Kramps, T.; Peter, O.; Brunner, E.; Nellen, D.; Froesch, B.; Chatterjee, S.; Murone, M.; Züllig, S.; Basler, K. Wnt/wingless Signaling requires BCL9/legless-mediated recruitment of pygopus to the nuclear β-catenin-TCF complex. *Cell* **2002**, *109*, 47–60. [CrossRef] [PubMed]
35. Thompson, B.; Townsley, F.; Rosin-Arbesfeld, R.; Musisi, H.; Bienz, M. A new nuclear component of the Wnt signalling pathway. *Nat. Cell Biol.* **2002**, *4*, 367–373. [CrossRef] [PubMed]
36. Wu, M.; Dong, H.; Xu, C.; Sun, M.; Gao, H.; Bu, F.; Chen, J. The Wnt-dependent and Wnt-independent functions of BCL9 in development, tumorigenesis, and immunity: Implications in therapeutic opportunities. *Genes Dis.* **2023**, *11*, 701–710. [CrossRef] [PubMed] [PubMed Central]
37. Tufail, M.; Wu, C. Wnt3a is a promising target in colorectal cancer. *Med. Oncol.* **2023**, *40*, 86. [CrossRef] [PubMed]
38. Voloshanenko, O.; Erdmann, G.; Dubash, T.D.; Augustin, I.; Metzig, M.; Moffa, G.; Hundsrucker, C.; Kerr, G.; Sandmann, T.; Anchang, B.; et al. Wnt secretion is required to maintain high levels of Wnt activity in colon cancer cells. *Nat. Commun.* **2013**, *4*, 2610. [CrossRef] [PubMed] [PubMed Central]
39. Qi, L.; Sun, B.; Liu, Z.; Cheng, R.; Li, Y.; Zhao, X. Wnt3a expression is associated with epithelial-mesenchymal transition and promotes colon cancer progression. *J. Exp. Clin. Cancer Res.* **2014**, *33*, 107. [CrossRef] [PubMed] [PubMed Central]
40. Ferrer-Mayorga, G.; Niell, N.; Cantero, R.; González-Sancho, J.M.; del Peso, L.; Muñoz, A.; Larriba, M.J. Vitamin D and Wnt3A have additive and partially overlapping modulatory effects on gene expression and phenotype in human colon fibroblasts. *Sci. Rep.* **2019**, *9*, 8085. [CrossRef] [PubMed] [PubMed Central]
41. A Torres, M.; A Yang-Snyder, J.; Purcell, S.M.; A DeMarais, A.; McGrew, L.L.; Moon, R.T. Activities of the Wnt-1 class of secreted signaling factors are antagonized by the Wnt-5A class and by a dominant negative cadherin in early Xenopus development. *J. Cell Biol.* **1996**, *133*, 1123–1137. [CrossRef]
42. Bauer, M.; Bénard, J.; Gaasterland, T.; Willert, K.; Cappellen, D. WNT5A Encodes Two Isoforms with Distinct Functions in Cancers. *PLoS ONE* **2013**, *8*, e80526. [CrossRef]
43. Asem, M.S.; Buechler, S.; Wates, R.B.; Miller, D.L.; Stack, M.S. Wnt5a Signaling in Cancer. *Cancers* **2016**, *8*, 79. [CrossRef] [PubMed]
44. Tufail, M.; Wu, C. WNT5A: A double-edged sword in colorectal cancer progression. *Mutat. Res. Mol. Mech. Mutagen.* **2023**, *792*, 108465. [CrossRef] [PubMed]
45. Ficari, F.; Cama, A.; Valanzano, R.; Curia, M.C.; Palmirotta, R.; Aceto, G.; Esposito, D.L.; Crognale, S.; Lombardi, A.; Messerini, L.; et al. APC gene mutations and colorectal adenomatosis in familial adenomatous polyposis. *Br. J. Cancer* **2000**, *82*, 348–353. [CrossRef] [PubMed] [PubMed Central]
46. Catalano, T.; D'amico, E.; Moscatello, C.; Di Marcantonio, M.C.; Ferrone, A.; Bologna, G.; Selvaggi, F.; Lanuti, P.; Cotellese, R.; Curia, M.C.; et al. Oxidative Distress Induces Wnt/β-Catenin Pathway Modulation in Colorectal Cancer Cells: Perspectives on APC Retained Functions. *Cancers* **2021**, *13*, 6045. [CrossRef] [PubMed] [PubMed Central]
47. Silviera, M.L.; Smith, B.P.; Powell, J.; Sapienza, C. Epigenetic differences in normal colon mucosa of cancer patients suggest altered dietary metabolic pathways. *Cancer Prev. Res.* **2012**, *5*, 374–384. [CrossRef] [PubMed] [PubMed Central]
48. Zhan, T.; Rindtorff, N.; Boutros, M. Wnt signaling in cancer. *Oncogene* **2017**, *36*, 1461–1473. [CrossRef]
49. Aoki, K.; Taketo, M.M. Adenomatous polyposis coli (APC): A multi-functional tumor suppressor gene. *J. Cell Sci.* **2007**, *120 Pt 19*, 3327–3335. [CrossRef] [PubMed]
50. Xu, D.; Yuan, L.; Liu, X.; Li, M.; Zhang, F.; Gu, X.; Zhang, D.; Yang, Y.; Cui, B.; Tong, J.; et al. EphB6 overexpression and *Apc* mutation together promote colorectal cancer. *Oncotarget* **2016**, *7*, 31111–31121. [CrossRef]
51. Behrens, J.; Von Kries, J.P.; Kühl, M.; Bruhn, L.; Wedlich, D.; Grosschedl, R.; Birchmeier, W. Functional interaction of β-catenin with the transcription factor LEF-1. *Nature* **1996**, *382*, 638–642. [CrossRef] [PubMed]

52. Xiao, L.; Zhang, C.; Li, X.; Jia, C.; Chen, L.; Yuan, Y.; Gao, Q.; Lu, Z.; Feng, Y.; Zhao, R.; et al. LEF1 Enhances the Progression of Colonic Adenocarcinoma via Remodeling the Cell Motility Associated Structures. *Int. J. Mol. Sci.* **2021**, *22*, 10870. [CrossRef]
53. Clevers, H. Wnt/β-Catenin Signaling in Development and Disease. *Cell* **2006**, *127*, 469–480. [CrossRef] [PubMed]
54. Gao, X.; Mi, Y.; Ma, Y.; Jin, W. LEF1 regulates glioblastoma cell proliferation, migration, invasion, and cancer stem-like cell self-renewal. *Tumor Biol.* **2014**, *35*, 11505–11511. [CrossRef]
55. Hao, Y.-H.; Lafita-Navarro, M.C.; Zacharias, L.; Borenstein-Auerbach, N.; Kim, M.; Barnes, S.; Kim, J.; Shay, J.; DeBerardinis, R.J.; Conacci-Sorrell, M. Induction of LEF1 by MYC activates the WNT pathway and maintains cell proliferation. *Cell Commun. Signal.* **2019**, *17*, 129. [CrossRef] [PubMed]
56. Xie, Q.; Tang, T.; Pang, J.; Xu, J.; Yang, X.; Wang, L.; Huang, Y.; Huang, Z.; Liu, G.; Tong, D.; et al. LSD1 Promotes Bladder Cancer Progression by Upregulating LEF1 and Enhancing EMT. *Front. Oncol.* **2020**, *10*, 1234. [CrossRef] [PubMed] [PubMed Central]
57. Yuan, M.; Zhang, X.; Zhang, J.; Wang, K.; Zhang, Y.; Shang, W.; Zhang, Y.; Cui, J.; Shi, X.; Na, H.; et al. DC-SIGN-LEF1/TCF1-miR-185 feedback loop promotes colorectal cancer invasion and metastasis. *Cell Death Differ.* **2020**, *27*, 379–395. [CrossRef] [PubMed]
58. Kim, G.-H.; Fang, X.-Q.; Lim, W.-J.; Park, J.; Kang, T.-B.; Kim, J.H.; Lim, J.-H. Cinobufagin Suppresses Melanoma Cell Growth by Inhibiting LEF1. *Int. J. Mol. Sci.* **2020**, *21*, 6706. [CrossRef] [PubMed]
59. Blazquez, R.; Rietkötter, E.; Wenske, B.; Wlochowitz, D.; Sparrer, D.; Vollmer, E.; Müller, G.; Seegerer, J.; Sun, X.; Dettmer, K.; et al. LEF1 supports metastatic brain colonization by regulating glutathione metabolism and increasing ROS resistance in breast cancer. *Int. J. Cancer* **2020**, *146*, 3170–3183. [CrossRef] [PubMed]
60. Heino, S.; Fang, S.; Lähde, M.; Högström, J.; Nassiri, S.; Campbell, A.; Flanagan, D.; Raven, A.; Hodder, M.; Nasreddin, N.; et al. Lef1 restricts ectopic crypt formation and tumor cell growth in intestinal adenomas. *Sci. Adv.* **2021**, *7*, eabj0512. [CrossRef] [PubMed] [PubMed Central]
61. Sampietro, J.; Dahlberg, C.L.; Cho, U.S.; Hinds, T.R.; Kimelman, D.; Xu, W. Crystal Structure of a β-Catenin/BCL9/Tcf4 Complex. *Mol. Cell* **2006**, *24*, 293–300. [CrossRef] [PubMed]
62. Habib, S.J.; Acebrón, S.P. Wnt signalling in cell division: From mechanisms to tissue engineering. *Trends Cell Biol.* **2022**, *32*, 1035–1048. [CrossRef] [PubMed]
63. Chen, J.; Rajasekaran, M.; Xia, H.; Kong, S.N.; Deivasigamani, A.; Sekar, K.; Gao, H.; Swa, H.L.; Gunaratne, J.; Ooi, L.L.; et al. CDK 1-mediated BCL 9 phosphorylation inhibits clathrin to promote mitotic Wnt signalling. *EMBO J.* **2018**, *37*, e99395. [CrossRef] [PubMed] [PubMed Central]
64. Suzuki, K.; Okuno, Y.; Kawashima, N.; Muramatsu, H.; Okuno, T.; Wang, X.; Kataoka, S.; Sekiya, Y.; Hamada, M.; Murakami, N.; et al. *MEF2D-BCL9* Fusion Gene Is Associated with High-Risk Acute B-Cell Precursor Lymphoblastic Leukemia in Adolescents. *J. Clin. Oncol.* **2016**, *34*, 3451–3459. [CrossRef] [PubMed]
65. Feng, M.; Wu, Z.; Zhou, Y.; Wei, Z.; Tian, E.; Mei, S.; Zhu, Y.; Liu, C.; He, F.; Li, H.; et al. BCL9 regulates CD226 and CD96 checkpoints in CD8+ T cells to improve PD-1 response in cancer. *Signal Transduct. Target. Ther.* **2021**, *6*, 313. [CrossRef] [PubMed] [PubMed Central]
66. Kaur, N.; Chettiar, S.; Rathod, S.; Rath, P.; Muzumdar, D.; Shaikh, M.; Shiras, A. Wnt3a mediated activation of Wnt/β-catenin signaling promotes tumor progression in glioblastoma. *Mol. Cell. Neurosci.* **2013**, *54*, 44–57. [CrossRef] [PubMed]
67. Lamb, R.; Ablett, M.P.; Spence, K.; Landberg, G.; Sims, A.H.; Clarke, R.B. Wnt pathway activity in breast cancer sub-types and stem-like cells. *PLoS ONE* **2013**, *8*, e67811. [CrossRef] [PubMed] [PubMed Central]
68. Verras, M.; Brown, J.; Li, X.; Nusse, R.; Sun, Z. Wnt3a growth factor induces androgen receptor-mediated transcription and enhances cell growth in human prostate cancer cells. *Cancer Res.* **2004**, *64*, 8860–8866. [CrossRef] [PubMed]
69. Fox, S.A.; Richards, A.K.; Kusumah, I.; Perumal, V.; Bolitho, E.M.; Mutsaers, S.E.; Dharmarajan, A.M. Expression profile and function of Wnt signaling mechanisms in malignant mesothelioma cells. *Biochem. Biophys. Res. Commun.* **2013**, *440*, 82–87. [CrossRef] [PubMed]
70. Qiang, Y.-W.; Shaughnessy, J.D., Jr.; Yaccoby, S. Wnt3a signaling within bone inhibits multiple myeloma bone disease and tumor growth. *Blood* **2008**, *112*, 374–382. [CrossRef] [PubMed] [PubMed Central]
71. Nygren, M.K.; Døsen, G.; Hystad, M.E.; Stubberud, H.; Funderud, S.; Rian, E. Wnt3A activates canonical Wnt signalling in acute lymphoblastic leukaemia (ALL) cells and inhibits the proliferation of B-ALL cell lines. *Br. J. Haematol.* **2006**, *136*, 400–413. [CrossRef] [PubMed]
72. Lee, M.A.; Park, J.-H.; Rhyu, S.Y.; Oh, S.-T.; Kang, W.-K.; Kim, H.-N. Wnt3a expression is associated with MMP-9 expression in primary tumor and metastatic site in recurrent or stage IV colorectal cancer. *BMC Cancer* **2014**, *14*, 125. [CrossRef] [PubMed] [PubMed Central]
73. Sun, G.; Wu, L.; Sun, G.; Shi, X.; Cao, H.; Tang, W. WNT5a in Colorectal Cancer: Research Progress and Challenges. *Cancer Manag. Res.* **2021**, *13*, 2483–2498. [CrossRef] [PubMed]
74. Smith, K.; Bui, T.D.; Poulsom, R.; Kaklamanis, L.; Williams, G.; Harris, A.L. Up-regulation of macrophage wnt gene expression in adenoma-carcinoma progression of human colorectal cancer. *Br. J. Cancer* **1999**, *81*, 496–502. [CrossRef] [PubMed] [PubMed Central]
75. Ying, J.; Li, H.; Yu, J.; Ng, K.M.; Poon, F.F.; Wong, S.C.C.; Chan, A.T.; Sung, J.J.; Tao, Q. WNT5A exhibits tumor-suppressive activity through antagonizing the Wnt/β-catenin signaling, and is frequently methylated in colorectal cancer. *Clin. Cancer Res.* **2008**, *14*, 55–61. [CrossRef] [PubMed]

76. Reya, T.; Clevers, H. Wnt signalling in stem cells and cancer. *Nature* **2005**, *434*, 843–850. [CrossRef] [PubMed]
77. Abdelmaksoud-Dammak, R.; Miladi-Abdennadher, I.; Saadallah-Kallel, A.; Khabir, A.; Sellami-Boudawara, T.; Frikha, M.; Daoud, J.; Mokdad-Gargouri, R. Downregulation of WIF-1 and Wnt5a in patients with colorectal carcinoma: Clinical significance. *Tumor Biol.* **2014**, *35*, 7975–7982. [CrossRef] [PubMed]
78. Hibi, K.; Mizukami, H.; Goto, T.; Kitamura, Y.; Sakata, M.; Saito, M.; Ishibashi, K.; Kigawa, G.; Nemoto, H.; Sanada, Y. WNT5A gene is aberrantly methylated from the early stages of colorectal cancers. *Hepatogastroenterology* **2009**, *56*, 1007–1009. [PubMed]
79. Cao, Y.-C.; Yang, F.; Liu, X.-H.; Xin, X.; Wang, C.-C.; Geng, M. Expression of Wnt5a, APC, β-catenin and their clinical significance in human colorectal adenocarcinoma. *Zhonghua Zhong Liu Za Zhi* **2012**, *34*, 674–678. [CrossRef] [PubMed]
80. Rawson, J.B.; Mrkonjic, M.; Daftary, D.; Dicks, E.; Buchanan, D.D.; Younghusband, H.B.; Parfrey, P.S.; Young, J.P.; Pollett, A.; Green, R.C.; et al. Promoter methylation of Wnt5a is associated with microsatellite instability and BRAF V600E mutation in two large populations of colorectal cancer patients. *Br. J. Cancer* **2011**, *104*, 1906–1912. [CrossRef] [PubMed] [PubMed Central]
81. Li, Q.; Chen, H. Silencing of *Wnt5a* during colon cancer metastasis involves histone modifications. *Epigenetics* **2012**, *7*, 551–558. [CrossRef] [PubMed] [PubMed Central]
82. Lai, C.; Robinson, J.; Clark, S.; Stamp, G.; Poulsom, R.; Silver, A. Elevation of *WNT5A* expression in polyp formation in *Lkb1*$^{+/-}$ mice and Peutz–Jeghers syndrome. *J. Pathol.* **2011**, *223*, 584–592. [CrossRef] [PubMed]
83. Bakker, E.R.; Das, A.M.; Helvensteijn, W.; Franken, P.F.; Swagemakers, S.; van der Valk, M.A.; Hagen, T.L.T.; Kuipers, E.J.; van Veelen, W.; Smits, R. Wnt5a promotes human colon cancer cell migration and invasion but does not augment intestinal tumorigenesis in Apc 1638N mice. *Carcinogenesis* **2013**, *34*, 2629–2638. [CrossRef] [PubMed]
84. Dong, X.; Liao, W.; Zhang, L.; Tu, X.; Hu, J.; Chen, T.; Dai, X.; Xiong, Y.; Liang, W.; Ding, C.; et al. RSPO2 suppresses colorectal cancer metastasis by counteracting the Wnt5a/Fzd7-driven noncanonical Wnt pathway. *Cancer Lett.* **2017**, *402*, 153–165. [CrossRef] [PubMed]
85. Chen, Z.; Tang, C.; Zhu, Y.; Xie, M.; He, D.; Pan, Q.; Zhang, P.; Hua, D.; Wang, T.; Jin, L.; et al. TrpC5 regulates differentiation through the Ca2+/Wnt5a signalling pathway in colorectal cancer. *Clin. Sci.* **2017**, *131*, 227–237. [CrossRef] [PubMed]
86. Schmitt, M.; Greten, F.R. The inflammatory pathogenesis of colorectal cancer. *Nat. Rev. Immunol.* **2021**, *21*, 653–667. [CrossRef] [PubMed]
87. Jridi, I.; Canté-Barrett, K.; Pike-Overzet, K.; Staal, F.J.T. Inflammation and Wnt Signaling: Target for Immunomodulatory Therapy? *Front. Cell Dev. Biol.* **2021**, *8*, 615131. [CrossRef]
88. Cui, G. TH9, TH17, and TH22 Cell Subsets and Their Main Cytokine Products in the Pathogenesis of Colorectal Cancer. *Front. Oncol.* **2019**, *9*, 1002. [CrossRef]
89. Liu, J.-L.; Yang, M.; Bai, J.-G.; Liu, Z.; Wang, X.-S. "Cold" colorectal cancer faces a bottleneck in immunotherapy. *World J. Gastrointest. Oncol.* **2023**, *15*, 240–250. [CrossRef]

biomedicines MDPI

Article

Demographic Characteristics and Survival in Young-Onset Colorectal Neuroendocrine Neoplasms

Deepak Vadehra [1], Sahithi Sonti [1], Beas Siromoni [2], Mrinalini Ramesh [3], Debduti Mukhopadhyay [3], Adrienne Groman [4], Renuka Iyer [1] and Sarbajit Mukherjee [1,*]

[1] Department of Medicine, Roswell Park Comprehensive Cancer Center, Buffalo, NY 14263, USA; deepak.vadehra@roswellpark.org (D.V.); sahithi.sonti@gmail.com (S.S.); renuka.iyer@roswellpark.org (R.I.)
[2] School of Health Sciences, University of South Dakota, Vermillion, SD 57069, USA; beassiromoni@gmail.com
[3] Department of Internal Medicine, University at Buffalo, Buffalo, NY 14204, USA; mramesh3@buffalo.edu (M.R.); debdutim@buffalo.edu (D.M.)
[4] Department of Biostatistics and Bioinformatics, Roswell Park Comprehensive Cancer Center, Buffalo, NY 14263, USA; adrienne.groman@roswellpark.org
* Correspondence: sarbajit.mukherjee@roswellpark.org; Tel.: +1-716-845-1300; Fax: +1-716-845-8935

Abstract: Background/Objectives: Recent epidemiological studies have revealed an upward trend in young-onset colorectal cancer (YOCRC) overall, whereas specific data on young-onset colorectal neuroendocrine neoplasms (YONEN) remain limited. This study investigated the demographic characteristics and survival trends in YONEN and compared these with those of young-onset colorectal adenocarcinoma (YOADC), the most common histologic subtype of YOCRC. Methods: A retrospective analysis was conducted from 2000 to 2019 using the Surveillance, Epidemiology, and End Results (SEER) database. Survival outcomes were assessed using univariate and multivariable Cox proportional models, with demographic differences evaluated via Wilcoxon rank sum and Chi-square tests. Results: Out of 61,705 patients aged 20–49 with colorectal cancer, 8% had NEN, and 92% had adenocarcinoma. The YONEN cohort had a higher proportion of Black patients and a lower proportion of White patients than the YOADC cohort (21% vs. 13% and 44% vs. 57%, respectively). NEN was more commonly found in the rectum (79%), and adenocarcinoma was mostly colonic (57%) in origin. YONEN patients had better survival than YOADC patients. Multivariate analysis in YONEN patients revealed that Hispanic patients had better overall survival compared to White patients (HR 0.67, 95% CI 0.47–0.95, $p = 0.024$). Conclusions: Racial disparities should be investigated further to aid in policymaking and targeted interventions.

Keywords: young-onset neuroendocrine neoplasms; young-onset adenocarcinoma; young-onset colorectal cancer (YOCRC)

1. Introduction

Colorectal cancer (CRC) is the second leading cause of cancer-related death in the United States, with adenocarcinoma comprising the majority. According to the American Cancer Society, an estimated 52,550 individuals will succumb to CRC in 2023 [1]. A recent study by Tan et al. examined the mortality trends in colorectal cancer in the US [2]. The study revealed that the incidence rates in CRC from 1999 to 2020 decreased significantly from 26.42 to 15.98 per 100,000 individuals, with an Average Annual Percent Change (AAPC) of −2.41. However, the Age-Adjusted Mortality Rate (AAMR) of rectosigmoid cancer went up from 0.82 to 1.08 per 100,000 individuals, with an AAPC of +1.10. Males and Black patients had the highest AAMRs, with rates of 23.90 and 26.93 per 100,000 individuals, respectively. Moreover, the overall AAMR of CRC decreased for those aged 50 years and older but worsened from 1.02 to 1.58 per 100,000 individuals for YOCRC patients, with an AAPC of +0.75. These results show that disparities in CRC mortality persist across age, sex, race, geographic region, and urbanization level, underscoring the necessity for targeted public health interventions.

A study which investigated the impact of race in receiving guideline-concordant care for young-onset colorectal cancer (YOCRC) in the United States revealed significant findings [3]. Black patients with YOCRC were more likely to be deprived of surgery (adjusted odds ratio [aOR] 1.15, 95% confidence interval [CI] 1.07 to 1.24), have standard (less than 12) lymph nodes examined (aOR 1.11, 95% CI 1.05 to 1.17), and not receive chemotherapy (aOR 1.22, 95% CI 1.17 to 1.27) compared to Caucasian patients. Black patients with rectal cancer were more likely not to have complete staging (aOR 1.90, 95% CI 1.77 to 2.04), not undergo surgery (aOR 1.38, 95% CI 1.30 to 1.45) or chemotherapy (aOR 1.68, 95% CI 1.56 to 1.82), not start radiotherapy (aOR 1.20, 95% CI 1.14 to 1.27), not finish radiotherapy (aOR 1.20, 95% CI 1.12 to 1.30), and be given treatment in the incorrect order (aOR 1.25, 95% CI 1.16 to 1.34).

The projected figures for young-onset colorectal cancer (YOCRC) reveal a troubling pattern within the 20 to 49-year-old age bracket. By 2040, colorectal cancer is anticipated to rank as the second most prevalent cancer in this demographic due to its increasing incidence [4]. Significantly, instances and fatalities related to colorectal cancer in younger adults have shown an upward trajectory over the past decade and are projected to continue increasing over the next two decades. This concerning trend may be attributed to factors such as sedentary lifestyles, poor dietary habits, obesity, and a lack of routine screening in this age group [4]. These forecasts emphasize the critical need for heightened awareness, early detection, and screening programs to tackle the escalating burden of colorectal cancer among younger cohorts.

From a histological perspective, adenocarcinoma is the most common subtype of CRC, followed by neuroendocrine neoplasms (NEN), which are a rare subgroup of young-onset colorectal cancers [5]. According to the WHO system, the grades for neuroendocrine neoplasms (NEN) include grade 1, grade 2, grade 3 which are distinguished by the mitotic rate and Ki-67 indices [6]. Tumors are further divided into neuroendocrine carcinomas (NEC), which are high-grade with further subdivision into multiple categories. One of these categories includes mixed adenoneuroendocrine carcinomas (MaNEC). Mixed neuroendocrine non-neuroendocrine neoplasms (MiNEN), on the other hand, are a rare group of NENs that consist of a neuroendocrine and a non-neuroendocrine component, both exceeding 30%. They can be either well or poorly differentiated and were included as a separate category of NENs in the 2019 WHO classification [6]. A recent study by Abboud et al. showed that the rise in YOCRC may be attributed to the fact that the incidence of colorectal NENs is increasing at a rate even faster than adenocarcinoma in the young population [7]. It was observed that there has been a substantial increase in the incidence of neuroendocrine neoplasms compared to adenocarcinomas (ADC) in this population. Specifically, the incidence of NENs showed a much more significant rise than that of ADC, with an average annual percentage change (AAPC) of 2.65 for NENs compared to 0.91 for ADC. This difference in AAPC between NENs and ADC was statistically significant ($p = 0.01$), indicating a notable disparity in the trends of these histopathological subtypes. These findings underscore the necessity for increased awareness and targeted screening strategies to address the rising incidence of colorectal cancer, particularly neuroendocrine neoplasms, in the younger population, ultimately aiming to enhance early detection and improve patient outcomes.

Similarly, Lumsdaine et al. conducted a population-based study from 1992 to 2015 that identified a substantial rise in young-onset colorectal cancer (CRC), specifically rectal neuroendocrine neoplasms (NENs) [8]. The incidence of rectal NENs exhibited a significant increase across all age groups, particularly notable in individuals aged 45–54 and those over 55 years. In the younger age brackets of 20 to 44 and 45 to 54 years, the annual percent changes (APCs) for rectal NENs were calculated at 2.9 and 6.1, respectively, indicating a notable upward trajectory in incidence rates. Notably, the surge in rectal NENs played a substantial role in the overall increase in rectal cancer cases, with statistics revealing that

NENs contributed significantly, accounting for 26.74% and 53.47% of the total increase in the respective age groups. These findings underscore the increasing impact of rectal NENs on the prevalence of young-onset colorectal cancer, underscoring the necessity for further research and clinical focus to address this emerging trend and its implications on patient care and management strategies.

Given the rarity of NENs and their typically slow progression, limited data exist concerning these tumors in younger individuals. Therefore, there is an unmet need to learn about overall survival (OS), disease-specific survival (DSS), and the factors affecting survival in young-onset colorectal NEN. This study aimed to explore patterns and disparities in survival rates among young individuals diagnosed with colorectal NEN and compare them with young-onset colorectal adenocarcinoma patients as well as those with average-onset colorectal NEN patients. Additionally, we aimed to identify factors affecting survival in colorectal NEN.

2. Materials and Methods

2.1. Study Design and Population

We conducted a retrospective study on YOCRC patients in the US between 2000 and 2019 using the Surveillance, Epidemiology, and End Results (SEER version 8.4.3 software, NIH/NCI, Bethesda, MD, USA) database. The SEER database compiles cancer-specific incidence data from population-based registries, covering approximately 35% of the US population [9]. Patients aged < 50 years were included. For the purpose of our analysis, we included adenocarcinomas, neuroendocrine tumors, neuroendocrine carcinomas, MiNEN, and MANEC. Patients with unknown stage, unknown grade, and unknown race were included in the model as they are SEER reportable statuses. Variables noted previously that could not be estimated were removed from the model. Our primary endpoint was to estimate overall survival (OS), defined as the time from diagnosis to death from any cause. Our secondary endpoint was disease-specific survival (DSS), defined as the time from diagnosis to death specifically due to cancer.

2.2. Covariates

Key variables of interest included the patient's demographic characteristics such as age, race, sex, and disease characteristics. Clinical variables of interest included disease stage, grade, surgery, primary cancer site, and year of diagnosis. Race was categorized into non-Hispanic Whites, non-Hispanic Blacks, Hispanics, Asian Americans, and Native Americans. The stage was classified according to the AJCC Classification System [10] as Stage 0, 1, 2, 3, and 4. Grade was classified as 1, 2, 3, and undifferentiated. The primary cancer site was categorized as colon, rectum, or rectosigmoid, and the year of diagnosis was divided into five periods: 2000–2003, 2004–2007, 2008–2011, 2012–2015, and 2016–2019.

2.3. Statistical Analysis

Demographic and clinical characteristics of the overall sample were summarized. For categorical variables, frequencies and relative frequencies were provided and compared using the Pearson chi-square test while contrasting YONENs and YOADCs. The Kruskal–Wallis test, a non-parametric test, was used for ordinal or continuous variables. The effects of various factors on OS and DSS were estimated using Cox proportional hazard models adjusted for age at diagnosis (20–39, 40–49, ≥50), sex, race, stage, grade, primary site, tumor size, surgery, and year of diagnosis. Result estimates were expressed as hazard ratios (HR) with 95% confidence intervals (CI). The Kaplan–Meier method calculated the OS and DSS. Yearly mortality rates per 100,000 were calculated to examine differences among histology and age groups. Unknown stage, unknown grade, and unknown race were included in the model as they are SEER reportable statuses. Variables that could not be estimated, were removed from the model. All model assumptions, proportional hazards and goodness of fit were evaluated visually using standard residual plots (Schoenfeld residuals versus time and standardized residuals versus predicted values). The significance level was denoted by $p < 0.05$. SAS, version 9.4 statistical software (SAS Institute Inc., Cary, NC, USA) was used for all statistical analyses.

3. Results

There were 61,705 patients in the young-onset colorectal cancer (YOCRC) group, of which 5128 belonged to the young-onset neuroendocrine neoplasms (YONEN) cohort, and the remaining 56,577 cases were part of the young-onset adenocarcinoma (YOADC) group. Fifty-two percent of the YONEN group were female, compared with 45.8% in the YOADC group. In the YONEN group, 43.6% were White, 20.6% were Black, 18.2% were Hispanic, 11.8% were Asian, and 1.2% were Native American. In the YOADC group, 58% were White, 17.6% Hispanic, 13.15% Black, 10.3% Asian, and 1% Native American. The primary disease site for most YONENs was the rectum (78.7%), followed by the colon (17.3%) and the rectosigmoid region (4%). Conversely, the majority of YOADCs had the colon as their primary disease site (61.2%), with 28.3% in the rectum and 10.5% in the rectosigmoid. The most common stage at diagnosis for YONENs was Stage I (21%), followed by 6.3% with Stage IV disease. However, a significant majority (67.2%) did not have a reported stage in the SEER database. For the YOADC group, 29% had Stage III disease at diagnosis, another 23.6% had Stage IV disease, with only 14.8% not having a reported stage. The demographic characteristics are described in Table 1.

Table 1. Demographic and clinical characteristics of young-onset colorectal neuroendocrine neoplasms and adenocarcinomas.

		Neuroendocrine Neoplasms	**Adenocarcinoma**
Age	20–39	1714 (33.4%)	14,164 (25%)
	40–49	3414 (66.6%)	42,413 (75%)
Sex	Male	2456 (47.9%)	33,108 (53.7%)
	Female	2672 (52.1%)	28,597 (46.3%)
Race	White	2235 (43.6%)	32,525 (57.5%)
	Black	1054 (20.6%)	7429 (13.1%)
	Asian	603 (11.8%)	5830 (10.3%)
	Native American	61 (1.2%)	573 (1.0%)
	Hispanic	935 (18.2%)	9937 (17.6%)
	Not reported	240 (4.7%)	283 (0.5%)
Year of diagnosis	2000–2003	794 (15.5%)	9748 (17.2%)
	2004–2007	950 (18.5%)	10,911 (19.3%)
	2008–2011	1054 (20.6%)	11,129 (19.7%)
	2012–2015	1122 (21.9%)	11,545 (20.4%)
	2016–2019	1208 (23.6%)	13,244 (23.4%)
Disease Site	Colon	887 (17.3%)	34,614 (61.2%)
	Rectosigmoid	204 (4.0%)	5954 (10.5%)
	Rectum	4037 (78.7%)	16,009 (28.3%)
Disease Grade	I	1760 (34.3%)	3848 (6.8%)
	II	302 (5.9%)	36,234 (64%)
	III	220 (4.3%)	9405 (16.6%)
	Undifferentiated	112 (2.2%)	971 (1.7%)
	Not reported	2734 (53.3%)	6119 (10.8%)
Overall Stage	I	1076 (21.0%)	6873 (12.1%)
	II	102 (2.0%)	10,695 (18.9%)
	III	129 (2.5%)	16,526 (29.2%)
	IV	323 (6.3%)	13,349 (23.6%)
	Not reported	3448 (67.2%)	8372 (14.8%)

We also compared young and average-onset NEN patients. We found that of patients from the younger group (<50 years), 52% of the group were female patients, compared with 48.9% in the average-onset group (patients aged 50 and above). Hispanic patients were overrepresented in the younger age group (18.2%), compared to the average-onset group (13.5%). YONEN patients had a higher proportion of rectal tumors compared to the average-onset NEN patients (79% vs. 69%). Patients in the average-onset group were also found to have a significantly higher proportion of grade III tumors (8%) compared to YONEN patients (4.3%), $p < 0.001$. This comparison is detailed in Table 2.

Table 2. Colorectal neuroendocrine neoplasms: descriptive statistics by age (<50 vs. ≥50).

		<50	**≥50**	***p*-Value**
Sex	Male	2456 (47.9%)	10,648 (51.1%)	$p < 0.001$
	Female	2672 (52.1%)	10,209 (48.9%)	
Race	White	2235 (43.6%)	10,420 (50.0%)	$p < 0.001$
	Black	1054 (20.6%)	4214 (20.2%)	
	Asian	603 (11.8%)	2490 (11.9%)	
	Native American	61 (1.2%)	158 (0.8%)	
	Hispanic	935 (18.2%)	2825 (13.5%)	
	Not reported	240 (4.7%)	750 (3.6%)	
Year of diagnosis	2000–2003	794 (15.5%)	2773 (13.3%)	$p < 0.001$
	2004–2007	950 (18.5%)	3579 (17.2%)	
	2008–2011	1054 (20.6%)	4428 (21.2%)	
	2012–2015	1122 (21.9%)	4931 (23.6%)	
	2016–2019	1208 (23.6%)	5146 (24.7%)	
Disease Site	Colon	887 (17.3%)	5668 (27.2%)	$p < 0.001$
	Rectosigmoid	204 (4.0%)	864 (4.1%)	
	Rectum	4037 (78.7%)	14,325 (68.7%)	
Disease Grade	I	1760 (34.3%)	6973 (33.4%)	$p < 0.001$
	II	302 (5.9%)	1272 (6.1%)	
	III	220 (4.3%)	1678 (8.0%)	
	Undifferentiated	112 (2.2%)	684 (3.3%)	
	Not reported	2734 (53.3%)	10,250 (49.1%)	
Overall Stage	In situ	50 (1.0%)	192 (0.9%)	$p < 0.001$
	I	1076 (21.0%)	3945 (18.9%)	
	II	102 (2.0%)	473 (2.3%)	
	III	129 (2.5%)	1080 (5.2%)	
	IV	323 (6.3%)	1760 (8.4%)	
	Not reported	3448 (67.2%)	13,407 (64.3%)	
Tumor size in mm	Mean/StdErr	15.1/0. 5	20.3/0.4	$p < 0.001$

We analyzed overall survival and disease-specific survival for YONEN and YOADC patients. YONENs had a 1-year survival rate of 0.94 (95% CI: 0.93, 0.94) and a 5-year survival rate of 0.88 (95% CI: 0.87, 0.89), whereas YOADCs had a 1-year survival rate of 0.90 (95% CI: 0.90, 0.90) and a 5-year survival rate of 0.63 (95% CI: 0.63, 0.64) with a median follow up time of 105 months. The 1-year disease-specific survival (DSS) rate for YONENs was 0.95 (95% CI: 0.94, 0.95) compared to 0.91 (95% CI: 0.91, 0.91) for YOADCs. The 5-year DSS rate for YONENs was 0.91 (95% CI: 0.90, 0.92) versus 0.66 (95% CI: 0.65, 0.66) for YOADCs with a median follow up time of 97 months. Survival rates by age group are detailed in Table 3 as well as the Kaplan–Meier curves in Figures A1 and A2. Cumulative incidence function and the Fine–Gray model were also performed, and the factors found to be significantly associated with DSS in the Cox regression were also retained in the Fine–Gray model (Figure A3).

Next, we examined the mortality trends. The rate of deaths per 100,000 decreased from 49,554 in 2000–2003 to 15,527 in the period between 2016 and 2019. Despite the rising incidence, there was a numerical improvement in mortality rates in recent years; the YONEN mortality rate decreased from 20,403 per 100,000 in 2000–2003 to 6705.3 per 100,000 in 2016–2019 (Figure 1). A similar magnitude of decrease was observed in the YOADC group, from 51,928.6 per 100,000 in 2000–2003 to 16,331.9 in 2016–2019 ($p < 0.001$) (Figure 1).

Table 3. Age-specific OS and DSS rates at 1 year and 5 years in YONEN and YOADC patients.

		Age Group (Years)	
		20–39	**40–49**
Overall Survival	1-year survival rate (NEN) (95% CI) (N = 5128)	0.96 (0.94, 0.96)	0.93 (0.92, 0.94)
	5-years survival rate (NEN) (95% CI) (N = 5128)	0.91 (0.89, 0.92)	0.87 (0.86, 0.88)
	1-year survival rate (ADC) (95% CI) (N = 56,577)	0.90 (0.89, 0.90)	0.90 (0.90, 0.90)
	5-years survival rate (ADC) (95% CI) (N = 56,577)	0.62 (0.61, 0.63)	0.64 (0.63, 0.64)
Disease-specific survival	1-year disease-specific survival rate (NEN) (95% CI) (N = 5128)	0.96 (0.95, 0.97)	0.94 (0.93, 0.95)
	5-years disease-specific-survival rate (NEN) (95% CI) (N = 5128)	0.93 (0.91, 0.94)	0.90 (0.89, 0.91)
	1-year disease-specific-survival rate (ADC) (95% CI) (N = 56,577)	0.90 (0.90, 0.91)	0.91(0.91, 0.91)
	5-years disease-specific-survival rate (ADC) (95% CI) (N = 56,577)	0.64 (0.63, 0.65)	0.66 (0.66, 0.67)

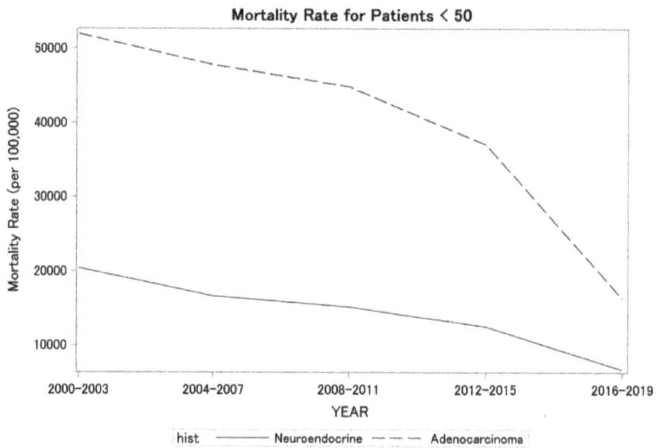

Figure 1. Mortality rates in YOADC and YONEN in the last two decades (per 100,000).

We also examined survival differences between the young-onset (<50) and average-onset (≥50) colorectal neuroendocrine neoplasms cohorts and found that the overall survival in the young-onset cohort was better than the average-onset NEN cohort. The 5-year overall survival rate between the young-onset and average-onset groups was 0.89 (95% CI: 0.88, 0.89) versus 0.65 (95% CI: 0.64, 0.66) after a median follow up time of 96 months. Similarly, the 5-year disease-specific survival rate for the young- versus average-onset groups was 0.92 (95% CI: 0.92, 0.92) versus 0.76 (95% CI: 0.75, 0.77) after a median follow up time of 82 months. This is likely related to the fact that younger patients have fewer comorbidities and better functional status, and hence can tolerate more aggressive treatments, leading to better survival rates. Survival rates of the young-onset versus average-onset colorectal NEN are detailed in Table 4.

Table 4. Age-specific OS and DSS rates between young-onset (<50) vs. average-onset (≥50) colorectal NENs at 1 Year and 5 Years.

Age Group (Years)	1-year Overall Survival Rate (NEN) (95% CI) (N = 25,985)	5-years Overall Survival Rate (NEN) (95% CI) (N = 25,985)	1-year Disease-Specific Survival Rate (NEN) (95% CI) (N = 25,985)	5-years Disease-Specific Survival Rate (NEN) (95% CI) (N = 25,985)
<50	0.94 (0.94, 0.95)	0.89 (0.88, 0.89)	0.95 (0.95, 0.96)	0.92 (0.92, 0.92)
≥50	0.80 (0.80, 0.81)	0.65 (0.64, 0.66)	0.85 (0.84, 0.85)	0.76 (0.75, 0.77)

Finally, we assessed factors affecting overall survival and disease-specific survival in the YONEN population. Higher grade was associated with worse overall survival (OS) (Grade III vs. Grade I; HR = 11.88, $p < 0.001$ [RMST Grade III: −1.2, CI −1.65, −0.75,

$p < 0.001$] and Undifferentiated vs. Grade I; HR = 11.23, $p < 0.001$ [RMST Undifferentiated: −1.65, CI −2.5, −0.79, $p < 0.001$]). The Hispanic race was also found to be associated with improved overall survival (HR = 0.67, $p = 0.024$ and RMST Hispanic: 0.042, CI 0.009, 0.074, $p = 0.013$). See Tables A1 and A2. Restricted mean survival time regression was applied on both OS and DSS, see Tables A3 and A4.

4. Discussion

The increase in colorectal cancer in young people is attributable to the rise in both neuroendocrine neoplasms and adenocarcinomas. However, Abboud et al. [7] have shown that the rate of increase in neuroendocrine neoplasms in the population under 50 years is significantly higher than that of adenocarcinoma. Adolescents and young adults (AYA) represent a unique population up to age 39 years of age. These patients are distinctive, as rare cancers are overrepresented in this group [11,12]. Given that this population does not routinely undergo screening colonoscopies, we wanted to see the distribution pattern of adenocarcinoma versus neuroendocrine neoplasms in these patients compared to those in the 40–49 age group. Similar to Abboud et al. [7], we noted that the AYA population is overrepresented with NEN compared to adenocarcinomas (Table 1).

We also noted several demographic differences between the YONEN and average-onset neuroendocrine neoplasm populations. When we compared these two groups, we found that female and Hispanic patients were overrepresented in the younger population and that the primary tumor site was more likely to be in the rectum. This calls for targeted interventions in younger female and Hispanic patients. Moreover, since these tumors are commonly found in the rectum, screening via flexible sigmoidoscopy should be considered in yearly testing.

A considerably larger fraction of YONEN patients were Black when compared to those with adenocarcinoma (20.6% vs. 13.1%) ($p < 0.001$). This is consistent with what has been seen in other studies. In a SEER analysis [13] of all NENs, it was seen that Black patients had a higher incidence and worse survival when compared to other races. However, our multivariate analysis did not identify Black race as an individual prognostic factor. This could be due to relatively small sample size in our study. In another study, Herring et al. showed significant differences in gene expression between Black and White pancreatic NEN (pNEN) patients, indicating potential disparities in tumor microenvironment that could affect outcomes [14]. RNA sequencing of pNENs from Black and White patients identified 372 markedly differentially expressed genes and 179 enriched gene sets, with key pathways associated with angiogenesis, blood vessel formation, cell migration, and immune response. Black patients showed enrichment in gene sets associated with blood vessel formation and cellular migration, while immune response pathways were downregulated in this group. These findings suggest distinct tumor biology in NENs from Black patients that may contribute to the disparate outcomes observed in this population, highlighting the importance of further validation and consideration of genetic ancestry in future studies.

Another critical difference we saw was the primary site of the tumor: 78.7% of YONENs were in the rectum, whereas 61.2% of YOADCs were primarily in the colon. Interestingly, we noted essential survival differences in the YONEN vs. YOADC population. Despite the rapid increase in the incidence of NENs, the median overall survival was high when compared to YOADCs. We looked at the overall survival rate in YONENs in the adolescent and young adult (AYA, 20–39 years) and 40–49-year-old subgroups. The age-specific OS rate slightly worsened with increasing age (0.96 vs. 0.93 1-year survival rate and 0.91 vs. 0.87 5-year survival rate, respectively). Nevertheless, this trend was not seen in the YOADCs (0.90 vs. 0.90 for a 1-year survival rate and 0.62 vs. 0.64 for a 5-year survival rate). It may be hypothesized that while the rise in incidence in YOCRC is attributable to the exponential increase in YONENs, the poor OS in this population is primarily still due to YOADCs, suggesting that adenocarcinoma in the young-onset cohort, specifically in the AYA cohort is an aggressive subtype [15,16].

Even with the differences in overall survival, mortality rates for both YOADC and YONENs consistently decreased over the last two decades in our study. This is in line with the decrease in mortality in all NENs, seen in a SEER analysis from 2017 [17] and an improved OS in the YOADC population seen in an NCDB analysis conducted in 2021 [18]. This may be attributable to multiple factors such as earlier screening and more treatment options in our therapeutic armamentarium. Research into available therapies is moving at a fast pace, with developments such as the tremendous success of PRRT [19,20].

Finally, we looked at the potential factors affecting mortality in the YONEN population and, predictably, found that the higher grade was associated with worse overall survival. We also found that Hispanic patients had better outcomes than White patients, corroborating the existing literature. A SEER analysis studying racial/ethnic disparities in non-pancreatic NENs found that Hispanic patients had the best overall survival when compared with non-Hispanic White and Black patients [21]. The study by Gosku et al. uncovered notable disparities in survival outcomes across racial and ethnic groups, shedding light on the nuanced impact of race and ethnicity on disease prognosis. Hispanic patients emerged as a cohort with distinct survival advantages, showcasing better overall survival rates than non-Hispanic White patients, with a hazard ratio of 0.89 (0.81–0.97). This lower risk of mortality among Hispanic individuals underscores a significant disparity in outcomes that warrants further investigation. Moreover, when examining specific primary tumor sites, Hispanic patients demonstrated superior overall survival in locations such as the small intestine and rectum, with hazard ratios of 0.81 (0.69–0.96) and 0.79 (0.63–0.99), respectively [20]. These findings suggest a potential biological or treatment-related advantage for Hispanic patients in these particular anatomical sites, highlighting the complexity of factors influencing survival disparities in neuroendocrine neoplasms.

The improved survival rates of Hispanics/Latinos compared to non-Hispanic Whites can be attributed to a complex interplay of genetic, behavioral, cultural, and environmental factors [22]. Genetic variances across racial/ethnic groups may influence survival outcomes. Behavioral differences, such as smoking patterns, also contribute, with Hispanics/Latinos potentially engaging in behaviors that confer a survival advantage. Despite often having lower socioeconomic status, this group may experience similar or better health outcomes, suggesting the influence of other factors. Cultural aspects like *familismo*, a strong family-oriented philosophy prevalent in Hispanic culture, could play a crucial role in promoting better health outcomes [22]. Additionally, survival advantages may arise not just from differences in healthcare access, but also from disparities in environmental exposures, cultural influences, and treatment approaches across racial/ethnic groups [22]. These elements together underscore the multifaceted reasons behind the superior survival outcomes observed in this demographic. Nevertheless, our findings highlight the critical role of race and ethnicity as independent prognostic factors in neuroendocrine neoplasms, emphasizing the importance of tailored interventions and personalized treatment strategies to address disparities and improve overall survival rates for diverse patient populations.

In moving forward from the findings of this study on young-onset colorectal neuroendocrine neoplasms, several key areas warrant further investigation to advance our understanding and improve patient outcomes. Better documentation at diagnosis of stage would aid in identifying differences in stage at presentation if any. Investigating genetic markers could lead to personalized treatment approaches, while examining the impact of lifestyle factors such as diet, exercise, and environmental exposures could provide valuable insights into disease development and progression. In addition, there are data showing that higher pain and stress scores using validated scales can correlate with poor outcomes in other diseases and the impact of these markers warrants further study in this group where anxiety and stress are expected to be high, underscoring the importance of databases that capture these measures [23]. Evaluating novel treatment modalities, including immunotherapy and targeted therapies, through randomized clinical trials, will help identify optimal strategies for YONEN patients.

Long-term follow-up studies are essential to assess survival outcomes and quality of life, helping to optimize patient care over time. Additionally, research on the impact of health insurance and care disparities is pivotal for addressing inequities, especially among minority groups [24,25]. Prior studies have documented disparities in relative survival that often impact minority groups [26]. Investigations have found that minority patients often have inadequate insurance that in turn results in increased risk of locally advanced disease on diagnosis [27]. Investigating barriers to healthcare access and designing interventions to reduce these disparities can improve outcomes [28]. Healthcare policy analysis should focus on evaluating existing policies and advocating for targeted interventions to reduce treatment disparities [29]. Implementing strategies like mobile screening programs and outreach oncology clinics can serve medically underserved communities, advancing health equity [30,31]. These research directions aim to tailor interventions to individual needs and enhance the quality of care and survival rates for YONEN patients.

Our study is not without limitations. Firstly, the study's retrospective design may introduce biases and limitations in data collection, analysis, and interpretation. For example, detection bias or misclassification bias could have impacted our survival outcomes. Secondly, we utilized data from the Surveillance, Epidemiology, and End Results (SEER) database, which may have limitations in terms of data accuracy, completeness, and consistency. The quality of the data in the SEER database, especially the significant amount of unreported data on disease stage and grade, might have impacted some of the study's findings like the multivariate analysis. Furthermore, the grades reported in the SEER database are not concordant with the WHO grading classification of NENs, with the database reporting Stage IV NEN [32]. Moreover, the classification of NETs has changed several times during the study period. We also removed the variables that could not be estimated from the multivariate model. Finally, the study may not have accounted for all potential confounding variables that could influence the outcomes of interest, such as comorbidities, or treatment variations like time to treatment initiation, quality of surgery, or type of chemotherapy, which were not available in the SEER database. Despite these limitations, our study was one of the largest studies to look at the demographic characteristics and survival of YONEN patients at a population level.

5. Conclusions

In conclusion, our study identified several sociodemographic disparities in YONEN and YOADC patients. We also found that despite increasing incidence rates, mortality rates have been steadily decreasing in this population. Further research is needed to understand the disparities for resource allocation as well as implementation of appropriate healthcare policies.

Author Contributions: D.V. and S.M. were involved in the conception of the study; D.V., S.S., S.M. and B.S. were involved in data curation and wrote the manuscript; A.G. was involved in data analysis and methodology; R.I., M.R. and D.M. reviewed and edited the manuscript. All authors have read and agreed to the published version of the manuscript.

Funding: This work was supported by funding from the National Cancer Institute (Grant number: P30CA016056). The study's design and decision to publish were independent of any involvement from the funding sources.

Institutional Review Board Statement: Not applicable.

Informed Consent Statement: Since de-identified retrospectively collected publicly available data were utilized, informed consent requirement was not required.

Data Availability Statement: All the data used for this study were from the publicly available SEER database.

Conflicts of Interest: Sarbajit Mukherjee serves as a volunteer guidelines panel member at the National Comprehensive Cancer Network and American Society of Clinical Oncology. He received research funding from the National Comprehensive Cancer Network and Ipsen Biopharmaceuticals/North American Neuroendocrine Tumor Society which were paid to the institute. Sarbajit Mukherjee received a consulting fee from Merck, Eisai, and Beigene. Sarbajit Mukherjee and all other authors declare no other conflicts of interest.

Abbreviations

YOAD	Young-onset adenocarcinoma
YONEN	Young-onset neuroendocrine neoplasms
SEER	Surveillance, Epidemiology, and End Results
OS	Overall survival
DSS	Disease-specific survival
CRC	Colorectal cancer
YOCRC	Young-Onset Colorectal Cancer
GEP-NENs	Gastroenteropancreatic neuroendocrine neoplasms
NCI	National Cancer Institute
RUCC	Rural/Urban Continuum Code
AJCC	American Joint Committee on Cancer
HR	Hazard Ratio
CI	Confidence Intervals
AYA	Adolescent and young adult
MiNEN	Mixed neuroendocrine non-neuroendocrine neoplasms
MANEC	Mixed adenoneuroendocrine carcinoma
AYA	Adolescents and young adults

Appendix A

Table A1. Factors affecting overall survival in YONEN (N = 2636).

Variable	Reference Group	Estimate (95% CL)	*p*-Value
Age			
40–49	20–39	1.14 (0.89, 1.46)	0.297
Gender			
Female	Male	0.87 (0.70, 1.08)	0.204
Grade			
II	I	1.41 (0.82, 2.43)	0.211
III	I	11.88 (8.16, 17.29)	<0.001
Undifferentiated	I	11.23 (7.37, 17.12)	<0.001
Not Reported	I	2.13 (1.48, 3.08)	<0.001
Race			
Asian	White	0.96 (0.65, 1.41)	0.842
Black	White	1.20 (0.91, 1.58)	0.203
Hispanic	White	0.67 (0.47, 0.95)	0.024
Native American	White	1.45 (0.53, 3.94)	0.471
Site			
Rectosigmoid	Colon	0.82 (0.47, 1.41)	0.468
Rectum	Colon	0.63 (0.45, 0.89)	0.008
Stage			
I	0	1.21 (0.17, 8.90)	0.849
II	0	1.32 (0.16, 10.58)	0.794
III	0	3.00 (0.40, 22.35)	0.283
IV	0	7.86 (1.07, 57.64)	0.043
Not Reported	0	1.75 (0.24, 12.70)	0.582
Surgical Procedure			
Colectomy	None	0.58 (0.40, 0.84)	0.003
Local Procedure/Wedge Resection	None	0.33 (0.24, 0.45)	<0.001
Total Proctectomy	None	0.31 (0.13, 0.71)	0.006
Total Proctocolectomy	None	0.42 (0.13, 1.36)	0.150
Year of Diagnosis			
2008–2011	2004–2007	0.92 (0.67, 1.26)	0.605
2012–2015	2004–2007	1.17 (0.83, 1.64)	0.371
2016–2019	2004–2007	1.33 (0.91, 1.95)	0.140

Table A2. Factors affecting disease-specific survival in YONEN (N = 2636).

Variable	Reference Group	Estimate (95% CL)	p-Value
Age			
40–49	20–39	1.16 (0.87, 1.54)	0.317
Gender			
Female	Male	0.78 (0.61, 1.00)	0.052
Grade			
II	I	2.57 (1.33, 4.99)	0.005
III	I	29.05 (18.12, 46.56)	<0.001
Undifferentiated	I	31.49 (18.96, 52.32)	<0.001
Not Reported	I	2.17 (1.33, 3.53)	0.002
Race			
Asian	White	1.15 (0.74, 1.78)	0.546
Black	White	1.02 (0.73, 1.43)	0.905
Hispanic	White	0.76 (0.52, 1.11)	0.155
Native American	White	0.49 (0.07, 3.57)	0.484
Site			
Rectosigmoid	Colon	1.00 (0.56, 1.77)	0.995
Rectum	Colon	0.45 (0.31, 0.65)	<0.001
Surgical Procedure			
Colectomy	None	0.40 (0.27, 0.58)	<0.001
Local Procedure/Wedge Resection	None	0.15 (0.10, 0.23)	<0.001
Procedure Not Reported	None	0.35 (0.05, 2.50)	0.294
Total Proctectomy	None	0.44 (0.19, 1.03)	0.059
Total Proctocolectomy	None	0.57 (0.18, 1.83)	0.342
Year of Diagnosis			
2008–2011	2004–2007	0.99 (0.69, 1.41)	0.954
2012–2015	2004–2007	1.04 (0.72, 1.51)	0.816
2016–2019	2004–2007	0.77 (0.51, 1.17)	0.220

Table A3. Restricted mean survival time model of OS in YONEN.

Variable	Estimate (95% CL)	p-Value
Age		
40–49	−0.017 (−0.042, 0.0085)	0.19
Gender		
Female	0.0095 (−0.017, 0.036)	0.49
Grade		
II	0.0094 (−0.042, 0.061)	0.72
III	−1.20 (−1.65, −0.75)	<0.001
Undifferentiated	−1.65 (−2.5, −0.79)	<0.001
Not Reported	−0.016 (−0.05, 0.018)	0.35
Race		
Asian	0.006 (−0.030, 0.0414)	0.74
Black	−0.028 (−0.068, 0.013)	0.19
Hispanic	0.042 (0.009, 0.074)	0.013
Native American	−0.034 (−0.15, 0.082)	0.56
Site		
Rectosigmoid	−0.0088 (−0.12, 0.10)	0.88
Rectum	0.036 (−0.046, 0.12)	0.39
Surgical Procedure		
Colectomy	0.16 (0.018, 0.29)	0.027
Local Procedure/Wedge Resection	0.13 (0.066, 0.20)	<0.001
Procedure Not Reported	0.11 (−0.16, 0.37)	0.43
Total Proctectomy	0.16 (−0.064, 0.38)	0.16
Total Proctocolectomy	0.23 (−0.0014, 0.45)	0.052
Year of Diagnosis		
2008–2011	0.02 (−0.039, 0.078)	0.51
2012–2015	0.015 (−0.044, 0.075)	0.61
2016–2019	−0.014 (−0.066, 0.038)	0.60

Table A4. Restricted mean survival time model of DSS in YONEN.

Variable	Estimate (95% CL)	*p*-Value
Age		
40–49	−0.01 (−0.03, 0.008)	0.26
Gender		
Female	−0.0043 (−0.023, 0.015)	0.66
Grade		
II	0.011 (−0.02, 0.042)	0.49
III	−0.005 (−0.047, 0.04)	0.83
Undifferentiated	−0.006 (−0.06, 0.04)	0.81
Not Reported	−0.0001 (−0.02, 0.02)	0.99
Race		
Asian	0.0065 (−0.02, 0.033)	0.62
Black	−0.024 (−0.054, 0.006)	0.12
Hispanic	0.024 (0.004, 0.04)	0.016
Native American	−0.053 (−0.17, 0.061)	0.36
Site		
Rectosigmoid	0.049 (−0.007, 0.11)	0.086
Rectum	0.022 (−0.029, 0.073)	0.39
Surgical Procedure		
Colectomy	0.014 (−0.05, 0.077)	0.68
Local Procedure/Wedge Resection	0.034 (−0.0023, 0.071)	0.067
Procedure Not Reported	−0.038 (−0.27, 0.2)	0.75
Total Proctectomy	0.048 (0.015, 0.082)	0.0048
Total Proctocolectomy	0.068 (0.026, 0.11)	0.0015
Year of Diagnosis		
2008–2011	−0.004 (−0.044, 0.037)	0.85
2012–2015	−0.011 (−0.05, 0.027)	0.56
2016–2019	−0.006 (−0.04, 0.028)	0.74

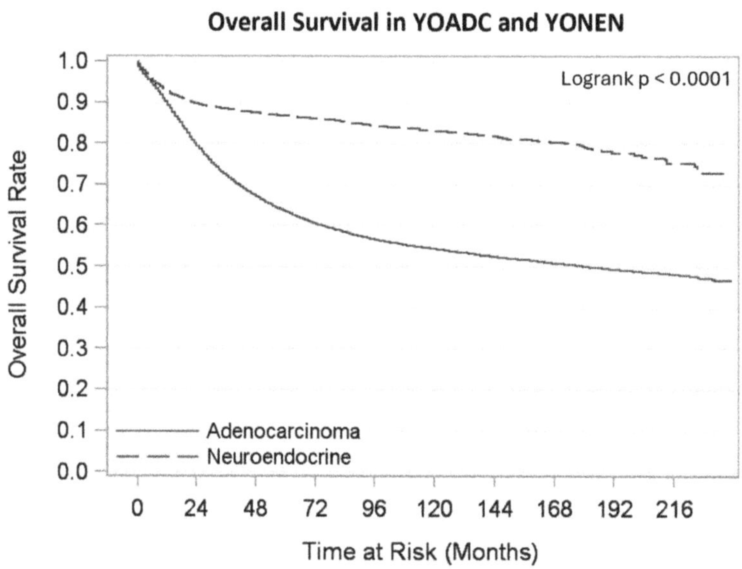

Figure A1. OS rates in YOADC and YONEN patients (N = 45,827).

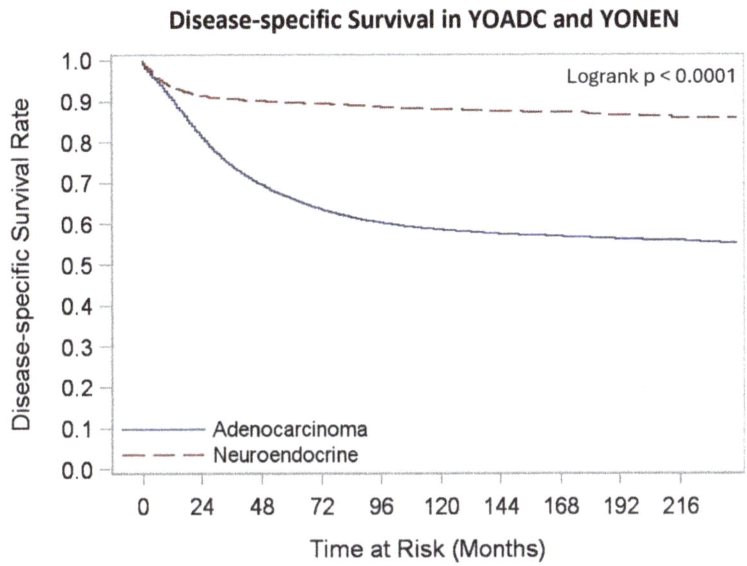

Figure A2. DSS rates in YOADC and YONEN patients (N = 45,827).

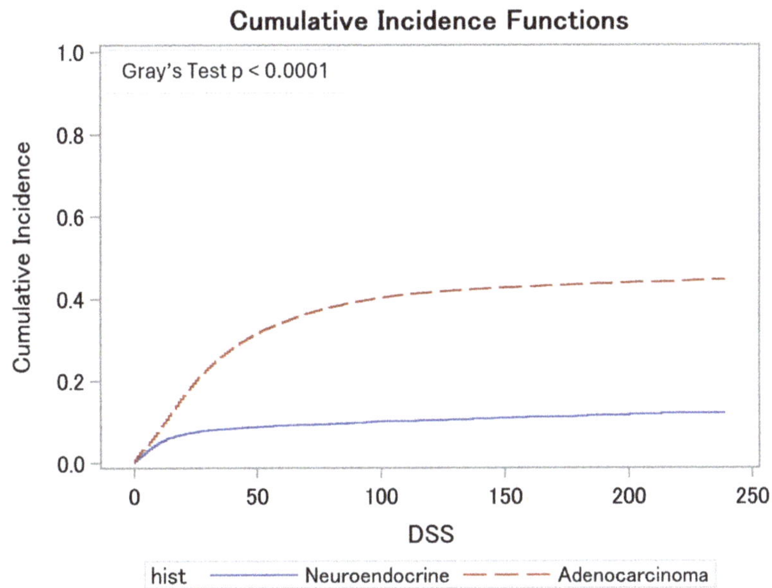

Figure A3. Cumulative incidence function and Fine–Gray model regarding DSS for YONEN versus YOADC.

References

1. Siegel, R.L.; Wagle, N.S.; Cercek, A.; Smith, R.A.; Jemal, A. Colorectal cancer statistics, 2023. *CA Cancer J. Clin.* **2023**, *73*, 233–254. [CrossRef] [PubMed]
2. Tan, J.Y.; Yeo, Y.H.; Ng, W.L.; Fong, Z.V.; Brady, J.T. How have US colorectal cancer mortality trends changed in the past 20 years? *Int. J. Cancer* **2024**, *155*, 493–500. [CrossRef] [PubMed]
3. Nogueira, L.M.; May, F.P.; Yabroff, K.R.; Siegel, R.L. Racial Disparities in Receipt of Guideline-Concordant Care for Early-Onset Colorectal Cancer in the United States. *J. Clin. Oncol.* **2024**, *42*, 1368–1377. [CrossRef]

4. Rahib, L.; Wehner, M.R.; Matrisian, L.M.; Nead, K.T. Estimated Projection of US Cancer Incidence and Death to 2040. *JAMA Netw. Open* **2021**, *4*, e214708. [CrossRef]
5. Stewart, S.L.; Wike, J.M.; Kato, I.; Lewis, D.R.; Michaud, F. A population-based study of colorectal cancer histology in the United States, 1998–2001. *Cancer* **2006**, *107* (Suppl. S5), 1128–1141. [CrossRef]
6. Helderman, N.C.; Suerink, M.; Kilinç, G.; van den Berg, J.G.; Nielsen, M.; Tesselaar, M.E.T. Relation between WHO Classification and Location- and Functionality-Based Classifications of Neuroendocrine Neoplasms of the Digestive Tract. *Neuroendocrinology* **2023**, *114*, 120–133. [CrossRef]
7. Abboud, Y.; Fraser, M.; Qureshi, I.; Hajifathalian, K. Early-Onset Colorectal Cancer: Are Neuroendocrine Tumors or Adenocarcinomas the Culprit? Analysis of the Largest U.S. Cancer Incidence Database, 2001–2020. *J. Clin. Med.* **2024**, *13*, 1098. [CrossRef]
8. Lumsdaine, C.T.; Liu-Smith, F.; Li, X.; Zell, J.A.; Lu, Y. Increased incidence of early onset colorectal adenocarcinoma is accompanied by an increased incidence of rectal neuroendocrine tumors. *Am. J. Cancer Res.* **2020**, *10*, 1888–1899.
9. Seer Incidence Data—Seer Data & Software. SEER. Available online: https://seer.cancer.gov/data/index.html (accessed on 29 April 2024).
10. Weiser, M.R. AJCC 8th Edition: Colorectal Cancer. *Ann. Surg. Oncol.* **2018**, *25*, 1454–1455. [CrossRef]
11. Barr, R.D.; Ferrari, A.; Ries, L.; Whelan, J.; Bleyer, W.A. Cancer in Adolescents and Young Adults: A Narrative Review of the Current Status and a View of the Future. *JAMA Pediatr.* **2016**, *170*, 495–501. [CrossRef]
12. Biller, L.H.; Ng, K. Pioneering a new care model for young-onset colorectal cancer: Innovations in clinical care and scientific discovery. *Color. Cancer* **2020**, *9*, CRC18. [CrossRef]
13. Shen, C.; Gu, D.; Zhou, S.; Xu, Y.; Sarshekeh, A.M.; Halperin, D.; Shih, Y.T.; Yao, J.C.; Dasari, A. Racial Differences in the Incidence and Survival of Patients with Neuroendocrine Tumors. *Pancreas* **2019**, *48*, 1373–1379. [CrossRef] [PubMed]
14. Herring, B.; Guenter, R.; Dhall, D.; Chen, H.; Yates, C.; Bart, R.J. Transcriptomic influences of racial disparities in black patients with pancreatic neuroendocrine tumors. *Endocr. Abstr.* **2022**, *89*, B12. [CrossRef]
15. Georgiou, A.; Khakoo, S.; Edwards, P.; Minchom, A.; Kouvelakis, K.; Kalaitzaki, E.; Nobar, N.; Calamai, V.; Ifijen, M.; Husson, O.; et al. Outcomes of Patients with Early Onset Colorectal Cancer Treated in a UK Specialist Cancer Center. *Cancers* **2019**, *11*, 1558. [CrossRef]
16. Sheth Bhutada, J.; Hwang, A.; Liu, L.; Deapen, D.; Freyer, D.R. Poor-Prognosis Metastatic Cancers in Adolescents and Young Adults: Incidence Patterns, Trends, and Disparities. *JNCI Cancer Spectr.* **2021**, *5*, pkab039. [CrossRef]
17. Dasari, A.; Shen, C.; Halperin, D.; Zhao, B.; Zhou, S.; Xu, Y.; Shih, T.; Yao, J.C. Trends in the Incidence, Prevalence, and Survival Outcomes in Patients with Neuroendocrine Tumors in the United States. *JAMA Oncol.* **2017**, *3*, 1335–1342. [CrossRef]
18. Cheng, E.; Blackburn, H.N.; Ng, K.; Spiegelman, D.; Irwin, M.L.; Ma, X.; Gross, C.P.; Tabung, F.K.; Giovannucci, E.L.; Kunz, P.L.; et al. Analysis of Survival Among Adults with Early-Onset Colorectal Cancer in the National Cancer Database. *JAMA Netw. Open* **2021**, *4*, e2112539. [CrossRef]
19. Singh, S.; Halperin, D.M.; Myrehaug, S.; Herrmann, K.; Pavel, M.; Kunz, P.L.; Chasen, B.; Capdevila, J.; Tafuto, S.; Oh, D.-Y.; et al. [177Lu]Lu-DOTA-TATE in newly diagnosed patients with advanced grade 2 and grade 3, well-differentiated gastroenteropancreatic neuroendocrine tumors: Primary analysis of the phase 3 randomized NETTER-2 study. *J. Clin. Oncol.* **2024**, *42* (Suppl. S3), LBA588. [CrossRef]
20. Strosberg, J.R.; Caplin, M.E.; Kunz, P.L.; Ruszniewski, P.B.; Bodei, L.; Hendifar, A.; Mittra, E.; Wolin, E.M.; Yao, J.C.; Pavel, M.E.; et al. ^{177}Lu-Dotatate plus long-acting octreotide versus high-dose long-acting octreotide in patients with midgut neuroendocrine tumours (NETTER-1): Final overall survival and long-term safety results from an open-label, randomised, controlled, phase 3 trial. *Lancet Oncol.* **2021**, *22*, 1752–1763. [CrossRef]
21. Goksu, S.Y.; Ozer, M.; Beg, M.S.; Sanford, N.N.; Ahn, C.; Fangman, B.D.; Goksu, B.B.; Verma, U.; Sanjeevaiah, A.; Hsiehchen, D.; et al. Racial/Ethnic Disparities and Survival Characteristics in Non-Pancreatic Gastrointestinal Tract Neuroendocrine Tumors. *Cancers* **2020**, *12*, 2990. [CrossRef]
22. Klugman, M.; Xue, X.; Ginsberg, M.; Cheng, H.; Rohan, T.; Hosgood, H.D., 3rd. Hispanics/Latinos in the Bronx Have Improved Survival in Non-Small Cell Lung Cancer Compared with Non-Hispanic Whites. *J. Racial Ethn. Health Disparities* **2020**, *7*, 316–326. [CrossRef] [PubMed]
23. Gandhi, S.; Ballman, K.V.; Holmes, M.D.; Visvanathan, K.; Symington, B.; Carvan, M.; Matyka, C.; Weiss, A.; Winer, E.P.; Razaq, W.; et al. Association of higher baseline stress and pain with clinical outcomes: Secondary analysis from Alliance A011502. *J. Clin. Oncol.* **2024**, *42* (Suppl. S16), 11016. [CrossRef]
24. Carethers, J.M. Racial and ethnic disparities in colorectal cancer incidence and mortality. *Adv. Cancer Res.* **2021**, *151*, 197–229. [CrossRef] [PubMed]
25. Hao, S.; Snyder, R.A.; Irish, W.; Parikh, A.A. Association of race and health insurance in treatment disparities of colon cancer: A retrospective analysis utilizing a national population database in the United States. *PLoS Med.* **2021**, *18*, e1003842. [CrossRef] [PubMed]
26. Zaki, T.A.; Liang, P.S.; May, F.P.; Murphy, C.C. Racial and Ethnic Disparities in Early-Onset Colorectal Cancer Survival. *Clin. Gastroenterol. Hepatol.* **2023**, *21*, 497–506.e3. [CrossRef]
27. Ko, N.Y.; Hong, S.; Winn, R.A.; Calip, G.S. Association of Insurance Status and Racial Disparities with the Detection of Early-Stage Breast Cancer. *JAMA Oncol.* **2020**, *6*, 385–392. [CrossRef]

28. Dickerson, J.C.; Ragavan, M.V.; Parikh, D.A.; Patel, M.I. Healthcare delivery interventions to reduce cancer disparities worldwide. *World J. Clin. Oncol.* **2020**, *11*, 705–722. [CrossRef]
29. Hollis, R.H.; Chu, D.I. Healthcare Disparities and Colorectal Cancer. *Surg. Oncol. Clin. N. Am.* **2022**, *31*, 157–169. [CrossRef]
30. Kale, S.; Hirani, S.; Vardhan, S.; Mishra, A.; Ghode, D.B.; Prasad, R.; Wanjari, M. Addressing Cancer Disparities Through Community Engagement: Lessons and Best Practices. *Cureus* **2023**, *15*, e43445. [CrossRef]
31. Richardson-Parry, A.; Baas, C.; Donde, S.; Ferraiolo, B.; Karmo, M.; Maravic, Z.; Münter, L.; Ricci-Cabello, I.; Silva, M.; Tinianov, S.; et al. Interventions to reduce cancer screening inequities: The perspective and role of patients, advocacy groups, and empowerment organizations. *Int. J. Equity Health* **2023**, *22*, 19. [CrossRef]
32. Punekar, S.R.; Kaakour, D.; Masri-Lavine, L.; Hajdu, C.; Newman, E.; Becker, D.J. Characterization of a Novel Entity of G3 (High-grade Well-differentiated) Colorectal Neuroendocrine Tumors (NET) in the SEER Database. *Am. J. Clin. Oncol.* **2020**, *43*, 846–849. [CrossRef]

biomedicines

MDPI

Article

Real-World Comparison of Trifluridine–Tipiracil with or Without Bevacizumab in Patients with Refractory Metastatic Colorectal Cancer

Hyunho Kim [1], Kabsoo Shin [2,3], Ho Jung An [1], In-Ho Kim [2,3], Jung Hoon Bae [4], Yoon Suk Lee [4], In Kyu Lee [4], MyungAh Lee [2,3] and Se Jun Park [2,3,*]

1 Division of Medical Oncology, Department of Internal Medicine, St. Vincent's Hospital, College of Medicine, The Catholic University of Korea, Suwon 16247, Republic of Korea; hori1104@naver.com (H.K.)
2 Division of Medical Oncology, Department of Internal Medicine, Seoul St. Mary's Hospital, College of Medicine, The Catholic University of Korea, Seoul 06591, Republic of Korea
3 Cancer Research Institute, College of Medicine, The Catholic University of Korea, Seoul 06591, Republic of Korea
4 Division of Colorectal Surgery, Department of Surgery, Seoul St. Mary's Hospital, College of Medicine, The Catholic University of Korea, Seoul 06591, Republic of Korea; eysi0815@catholic.ac.kr (J.H.B.); cmcgslee@catholic.ac.kr (I.K.L.)
* Correspondence: psj6936@naver.com; Tel.: +82-2-2258-6757

Abstract: Background/Objectives: Patients with metastatic colorectal cancer (mCRC) who are refractory to standard chemotherapy face limited treatment options. While trifluridine–tipiracil (FTD–TPI) and regorafenib have shown modest efficacy in prior clinical trials, recent data from the SUNLIGHT trial demonstrated that combining FTD–TPI with bevacizumab (FTD–TPI+BEV) may improve overall survival compared to FTD–TPI alone. However, supporting evidence from real-world populations remains scarce. **Methods:** This retrospective study assessed the real-world effectiveness and safety of FTD–TPI+BEV versus FTD–TPI monotherapy in patients with refractory mCRC treated at two institutions from June 2020 to October 2024. **Results:** A total of 106 patients were included, with 47 treated with FTD–TPI+BEV and 59 with FTD–TPI alone. Median progression-free survival (PFS) was significantly longer with FTD–TPI+BEV compared to FTD–TPI alone (4.1 vs. 2.1 months; HR = 0.56; $p = 0.004$), while median overall survival showed a non-significant trend favoring FTD–TPI+BEV (8.4 vs. 6.3 months; HR = 0.74; $p = 0.189$). The disease control rate was also significantly higher with FTD–TPI+BEV (59.6% vs. 25.4%, $p = 0.001$). Subgroup analyses showed consistent PFS benefits. Grade 3–5 adverse events occurred at comparable rates between groups. **Conclusions:** FTD–TPI+BEV may represent a preferred salvage treatment option for refractory mCRC.

Keywords: colorectal cancer; trifluridine–tipiracil; bevacizumab; real-world evidence

1. Introduction

Colorectal cancer (CRC) is the fourth most common cancer and the second leading cause of cancer-related mortality worldwide [1]. Approximately 50% of patients with colorectal cancer develop distant metastases over the course of the disease, with a 5-year overall survival rate of only 15% [2]. Standard systemic treatment for metastatic CRC (mCRC) typically consists of fluorouracil-based chemotherapy with oxaliplatin or irinotecan, combined with vascular endothelial growth factor (VEGF)-targeted treatments, such as bevacizumab or epidermal growth factor receptor (EGFR)-targeted therapies, which are used only in *RAS* wild-type tumors [2,3]. Recently, new targeted therapies have significantly improved outcomes in patients with mCRC harboring specific molecular alterations [4]. However, for patients with mCRC who fail standard cytotoxic combination therapy and lack actionable molecular alterations, effective treatment options remain limited. This highlights the urgent need to develop novel therapeutic strategies specifically for this subset of patients.

In the third-line or later treatment setting, trifluridine–tipiracil (FTD–TPI, also known as TAS-102) and regorafenib have demonstrated incremental improvements in median overall survival (OS) compared to placebo and are prescribed as salvage treatment options, as supported by the RECOURSE and CORRECT trials conducted in patients with mCRC refractory to standard chemotherapy [5,6]. More recently, the SUNLIGHT trial showed that FTD–TPI combined with bevacizumab (FTD–TPI+BEV) significantly prolonged survival compared to FTD–TPI alone, establishing it as the current preferred regimen for the treatment of refractory mCRC [7]. However, as the majority of patients in this study (93%) had received treatment only up to the second line and approximately 28% had not previously received anti-VEGF therapy, the applicability of these findings to real-world populations requires careful consideration due to potential differences in clinical characteristics.

A retrospective large-scale study using a claims database examined the effectiveness of FTD–TPI+BEV in real-world settings and demonstrated better OS outcomes compared to FTD–TPI alone or regorafenib [8]. Nevertheless, because the study relied on claims data, specific efficacy outcomes, such as progression-free survival (PFS) and response rate, could not be assessed. In addition, several meta-analyses have consistently suggested that FTD–TPI+BEV is associated with improved survival compared to FTD–TPI monotherapy in patients with refractory mCRC [9,10]. However, because these meta-analyses were based on trials with heterogeneous treatment arms and lacked clinicopathological data beyond survival outcomes, their applicability to real-world patient populations remains limited, making it difficult to identify which subgroups may benefit from the addition of bevacizumab to FTD–TPI.

For most patients with refractory mCRC without druggable molecular alterations, salvage treatments provide only modest effectiveness, and no standard regimen has been established. Given the differences in clinical characteristics between trial populations and real-world patients, along with various clinical factors influencing survival outcomes, evaluating the efficacy of combination therapies in real-world settings is essential. This study aims to assess the real-world effectiveness and safety of FTD–TPI+BEV compared to FTD–TPI alone as a third- or later-line treatment for refractory mCRC.

2. Materials and Methods

2.1. Patients

Patients with histologically confirmed mCRC were eligible for inclusion if they were refractory or intolerant to fluoropyrimidines, irinotecan, oxaliplatin, and an anti-EGFR monoclonal antibody (for *RAS* wild-type only). Prior exposure to bevacizumab, aflibercept, or regorafenib was permitted. This retrospective study reviewed the medical records of patients with mCRC who experienced treatment failure with standard cytotoxic chemotherapy at Seoul St. Mary's Hospital and St. Vincent's Hospital. The study was conducted in accordance with Korean regulatory requirements and the ethical principles outlined in the Declaration of Helsinki. The Institutional Review Board (IRB) of The Catholic University of Korea, Seoul St. Mary's Hospital, approved the study protocol (approval ID: KC25RISI0178) and granted a waiver of informed consent owing to its retrospective design.

2.2. Procedures

Patients received either FTD–TPI+BEV or FTD–TPI alone. FTD–TPI was administered orally at a dose of 35 mg/m^2 twice daily on days 1–5 and 8–12 of a 28-day cycle. Bevacizumab (5 mg/kg) was administered intravenously on days 1 and 15 of each 28-day cycle. Treatment was continued until radiological or clinical disease progression, unacceptable toxicities, or patient withdrawal. Treatment dose and schedule modifications were allowed to manage adverse events. Dose reductions for FTD–TPI were implemented in a stepwise manner according to the predefined protocol.

2.3. Assessments

Tumor evaluations were performed following the Response Evaluation Criteria in Solid Tumors (RECIST) version 1.1. Imaging data for these assessments were obtained through computed tomography or magnetic resonance imaging of the thorax, abdomen, and pelvis conducted at 8-week intervals from the initiation of chemotherapy. Additional imaging was carried out when clinically indicated. As part of the tumor's evaluation, carcinoembryonic antigen (CEA) levels were measured at baseline and every 8 weeks until disease progression. Adverse events were assessed at each clinic visit and graded based on the National Cancer Institute Common Terminology Criteria for Adverse Events, version 5.0.

2.4. Statistical Analysis

Descriptive statistics were presented as proportions for categorical variables and as medians with interquartile ranges (IQRs) for continuous variables. Comparisons of categorical variables were conducted using either the chi-square test or Fisher's exact test, while continuous variables were analyzed using Student's *t*-test. OS was defined as the duration from chemotherapy initiation to death from any cause, and PFS was defined as the time from chemotherapy initiation to disease progression or death, whichever occurred first. Patients who did not experience disease progression or death by the data cutoff date were censored, with follow-up duration calculated from chemotherapy initiation to the date of last clinical follow-up or data cutoff. Kaplan–Meier survival curves were generated for each treatment group to estimate median OS and PFS, and differences between groups were assessed using the unstratified log-rank test. Unstratified Cox proportional hazards regression was applied to determine hazard ratios (HRs) and their 95% confidence intervals (CIs). Additionally, a Cox regression model with forward stepwise selection was used to assess the impact of treatment and baseline prognostic factors on survival outcomes. The objective response rate was defined as the proportion of patients who achieved a best overall response of either complete or partial response. The disease control rate was defined as the proportion of patients achieving complete response, partial response, or stable disease, with stable disease required to persist for at least 6 weeks according to RECIST version 1.1 criteria. Fisher's exact test was used for pairwise comparisons of the objective response rate and the disease control rate between treatment groups. All statistical analyses were two-sided, with p-values < 0.05 considered statistically significant. Statistical analysis was performed using SPSS for Windows, version 24.0 (IBM SPSS Inc., Armonk, NY, USA), and GraphPad Prism, version 10.4 (GraphPad Software Inc., San Diego, CA, USA).

3. Results

3.1. Patients

Between June 2020 and October 2024, a total of 106 patients were included in the study, with 47 treated with FTD–TPI+BEV and 59 treated with FTD–TPI alone. Baseline characteristics of the study population are summarized in Table 1. The median age was 57 years

(IQR, 50–64), with no significant difference between the two groups. The proportion of females was higher in the FTD–TPI+BEV group than in the FTD–TPI group (66.0% vs. 40.7%, $p = 0.010$), and patients in the FTD–TPI+BEV group were more likely to have a better Eastern Cooperative Oncology Group (ECOG) performance status (ECOG 0–1: 83.0% vs. 59.3%, $p = 0.008$). A slightly higher proportion of patients in the FTD–TPI+BEV group had a metastatic disease duration of ≤18 months compared to those in the FTD–TPI group (34.0% vs. 18.6%, $p = 0.071$). A higher percentage of patients in the FTD–TPI+BEV group had received only two prior lines of therapy (78.7% vs. 59.3%, $p = 0.034$). Nevertheless, prior exposure to fluoropyrimidine, irinotecan, oxaliplatin, and anti-VEGF therapy, as well as anti-EGFR therapy (for *RAS* wild-type tumors only), was comparable between the two groups.

Table 1. Baseline characteristics.

Variable	Total (n = 106)	FTD–TPI+BEV (n = 47)	FTD–TPI (n = 59)	p Value
Age, years	57 (50–64)	55 (49–62)	59 (50–64)	0.324
≥65 yr	22 (20.8)	9 (19.1)	13 (22.0)	0.716
Gender				
Male	51 (48.1)	16 (34.0)	35 (59.3)	0.010
Female	55 (51.9)	31 (66.0)	24 (40.7)	
ECOG performance status				
0–1	74 (69.8)	39 (83.0)	35 (59.3)	0.008
2	32 (30.2)	8 (17.0)	24 (40.7)	
Primary diagnosis				
Colon cancer	65 (61.3)	31 (66.0)	34 (57.6)	0.382
Rectal cancer	41 (38.7)	16 (34.0)	25 (42.4)	
Primary tumor location				
Right side	16 (15.1)	11 (23.4)	5 (8.5)	0.054
Left side	90 (84.9)	36 (76.6)	54 (91.5)	
Histology				
Adenocarcinoma	99 (93.4)	46 (97.9)	53 (89.8)	0.129
Mucinous carcinoma	7 (6.6)	1 (2.1)	6 (10.2)	
Duration of metastatic disease				
Median, months	27.2 (17.9–49.3)	23.9 (17.3–51.9)	29.1 (19.2–48.4)	0.876
<18 months	27 (25.5)	16 (34.0)	11 (18.6)	0.071
≥18 months	79 (74.5)	31 (66.0)	48 (81.4)	
Number of metastatic organ sites				
1 or 2	46 (43.4)	21 (44.7)	25 (42.4)	0.812
≥3	60 (56.6)	26 (55.3)	34 (57.6)	
***RAS* mutation status**				
Wild-type	36 (34.0)	14 (29.8)	22 (37.3)	0.418
Mutant	70 (66.0)	33 (70.2)	37 (62.7)	
***BRAF* mutation status**				
Wild-type	106 (100)	47 (100)	59 (100)	
MMR and MSI status				
MMR proficient or MSI stable	106 (100)	47 (100)	59 (100)	
Previous lines of therapy *				
2	72 (67.9)	37 (78.7)	35 (59.3)	0.034
≥3	34 (40.7)	10 (21.3)	24 (40.7)	
Previous treatments				
Fluoropyrimidine	106 (100)	47 (100)	59 (100)	
Irinotecan	106 (100)	47 (100)	59 (100)	
Oxaliplatin	106 (100)	47 (100)	59 (100)	
Anti-VEGF therapy	103 (97.2)	45 (95.7)	58 (98.3)	
Anti-EGFR therapy †	35 (97.2)	14 (100)	21 (95.4)	

Table 1. *Cont.*

Variable	Total (n = 106)	FTD–TPI+BEV (n = 47)	FTD–TPI (n = 59)	p Value
Neutrophil to lymphocyte ratio				
<3	43 (40.6)	18 (38.3)	25 (42.4)	0.671
≥3	63 (59.4)	29 (61.7)	34 (57.6)	
Baseline CEA				
<50 μg/L	42 (39.6)	22 (46.8)	20 (33.9)	0.177
≥50 μg/L	64 (60.4)	25 (53.2)	39 (66.1)	

FTD–TPI+BEV, trifluridine–tipiracil plus bevacizumab; FTD–TPI, trifluridine–tipiracil; ECOG, Eastern Cooperative Oncology Group; MMR, mismatch repair; MSI, microsatellite instability; VEGF, vascular endothelial growth factor; EGFR, epidermal growth factor receptor; CEA, carcinoembryonic antigen. Data are n (%) or median (IQR). * Systemic treatment for metastatic disease, including cytotoxic chemotherapy, targeted therapy, and receptor tyrosine kinase inhibitors. † Proportion of patients with *RAS* wild-type disease.

3.2. *Effectiveness*

The median follow-up duration was 7.3 months (95% CI, 5.6–8.4) based on the reverse Kaplan–Meier method. Among the 106 patients, disease progression was observed in 96 (90.6%), with the remaining 10 patients (9.4%) censored at the time of analysis. For overall survival, 85 patients (80.2%) had died, and the remaining 21 patients (19.8%) were censored. The median PFS for the entire cohort was 2.3 months (95% CI, 1.9–2.7), and the median OS was 7.3 months (95% CI, 5.9–8.7) (Figure S1). By treatment group, the median PFS was 4.1 months (95% CI, 2.8–5.3) in the FTD–TPI+BEV group and 2.1 months (95% CI, 1.9–2.3) in the FTD–TPI group (HR = 0.56, 95% CI, 0.37–0.83; p = 0.004, Figure 1A), demonstrating a significantly longer PFS with FTD–TPI+BEV. The median OS was 8.4 months (95% CI, 6.6–10.3) in the FTD–TPI+BEV group and 6.3 months (95% CI, 3.7–8.9) in the FTD–TPI group (HR = 0.74, 95% CI, 0.48–1.15; p = 0.189, Figure 1B). Although the FTD–TPI+BEV group exhibited a numerically longer OS, the difference was not statistically significant. Effectiveness outcomes according to treatment regimen are summarized in Table 2. The objective response rates were 2.1% in the FTD–TPI+BEV group and 1.7% in the FTD–TPI group, with no statistically significant difference between the two groups (p = 1.000), while the disease control rate was significantly higher in the FTD–TPI+BEV group than in the FTD–TPI group (59.6% vs. 25.4%, p = 0.001).

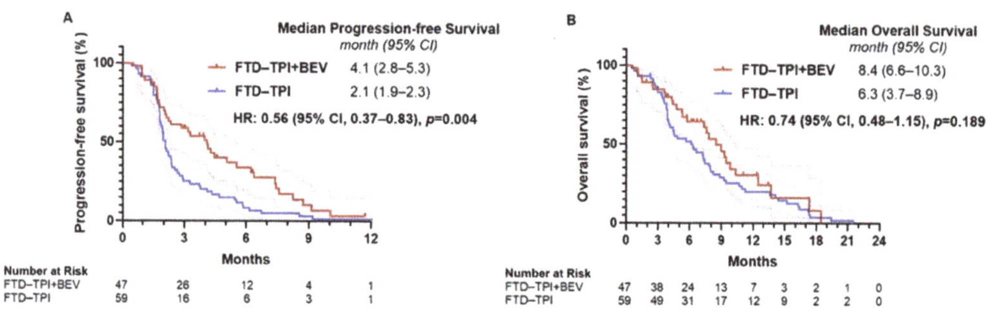

Figure 1. Kaplan–Meier estimates of (**A**) progression-free survival and (**B**) overall survival by treatment group.

Table 2. Effectiveness outcomes of trifluridine–tipiracil with bevacizumab compared to trifluridine–tipiracil alone in patients with metastatic colorectal cancer.

Variable	FTD–TPI+BEV (n = 47)	FTD–TPI (n = 59)	p Value
Best overall response, n (%)			
Partial response	1 (2.1)	1 (1.7)	
Stable disease	27 (57.4)	14 (23.7)	
Progressive disease	19 (40.4)	44 (74.6)	
Objective response rate, n (%)	1 (2.1)	1 (1.7)	1.000
Disease control rate, n (%)	28 (59.6)	15 (25.4)	0.001
Median PFS, months [95% CI]	4.1 [2.8–5.3]	2.1 [1.9–2.3]	
16-week PFS, % [95% CI]	51.0 [35.5–64.6]	20.3 [11.2–31.3]	
Median OS, months [95% CI]	8.4 [6.6–10.3]	6.3 [3.7–8.9]	
9-month OS, % [95% CI]	45.7 [28.1–61.7]	29.0 [17.9–41.0]	

FTD–TPI+BEV, trifluridine–tipiracil plus bevacizumab; FTD–TPI, trifluridine–tipiracil; PFS, progression-free survival; OS, overall survival.

3.3. Univariable and Multivariable Analysis for Survival Outcomes

Table 3 presents the findings from the univariable and multivariable analyses conducted to identify predictors of survival outcomes in the entire cohort, including patients treated with either FTD–TPI+BEV or FTD–TPI alone. In the multivariable analysis, peritoneal metastases (HR = 1.66, 95% CI, 1.07–2.58; p = 0.023) and a higher baseline neutrophil-to-lymphocyte ratio (NLR) (HR = 1.60, 95% CI, 1.03–2.48; p = 0.038) were significantly associated with shorter PFS. Additionally, poor performance status (HR = 1.56, 95% CI, 0.96–2.53; p = 0.071) showed a trend toward an association with shorter PFS. Treatment with FTD–TPI+BEV was significantly associated with longer PFS compared to FTD–TPI alone (HR = 0.60, 95% CI, 0.39–0.94; p = 0.025). Regarding OS, poor performance status (HR = 3.12, 95% CI, 1.88–5.19; p < 0.001), higher baseline NLR (HR = 1.64, 95% CI, 1.04–2.58; p = 0.032), and higher baseline CEA (HR = 1.93, 95% CI, 1.18–3.18; p = 0.009) were significantly associated with worse OS outcomes.

Table 3. Univariable and multivariable analyses of clinicopathologic features associated with progression-free survival and overall survival in patients with metastatic colorectal cancer.

Variables	PFS				OS			
	Univariable Analysis		Multivariable Analysis		Univariable Analysis		Multivariable Analysis	
	HR (95% CI)	p Value	HR (95% CI)	p Value	HR (95% CI)	p Value	HR (95% CI)	p Value
Age ≥ 65 yr (vs. <65 yr)	0.84 (0.51–1.38)	0.498			1.11 (0.61–2.01)	0.734		
Female (vs. male)	0.84 (0.56–1.27)	0.413			1.17 (0.76–1.80)	0.483		
ECOG PS 2 (vs. PS 0–1)	2.11 (1.36–3.27)	0.001	1.56 (0.96–2.53)	0.071	3.30 (2.06–5.29)	<0.001	3.12 (1.88–5.19)	<0.001
Rectal (vs. colon)	1.31 (0.87–1.98)	0.198			1.26 (0.81–1.96)	0.306		
Right side (vs. left side)	0.78 (0.44–1.38)	0.393			0.70 (0.36–1.36)	0.294		
RAS mutant (vs. wild-type)	1.02 (0.67–1.56)	0.934			1.13 (0.72–1.77)	0.600		
Previous lines of therapy >2 (vs. ≤2)	1.31 (0.84–2.06)	0.231			1.38 (0.88–2.15)	0.158		
Metastatic duration (<18 vs. ≥18 months)	1.05 (0.67–1.66)	0.833			0.92 (0.56–1.51)	0.735		
No. of sites of metastasis (≥3 vs. 0–2)	1.25 (0.83–1.88)	0.281			1.50 (0.96–2.34)	0.072		
Liver metastases (vs. none)	1.44 (0.93–2.24)	0.105			0.96 (0.61–1.50)	0.844		
Lung metastases (vs. none)	0.90 (0.57–1.40)	0.629			1.22 (0.75–1.99)	0.415		
Peritoneum metastases (vs. none)	1.71 (1.12–2.61)	0.013	1.66 (1.07–2.58)	0.023	1.24 (0.80–1.92)	0.342		
Neutrophil–lymphocyte ratio ≥3 (vs. <3)	1.63 (1.08–2.48)	0.022	1.60 (1.03–2.48)	0.038	1.79 (1.15–2.78)	0.010	1.64 (1.04–2.58)	0.032
CEA ≥50 μg/L (vs. <50 μg/L)	1.56 (1.03–2.37)	0.038	1.21 (0.77–1.91)	0.406	2.41 (1.50–3.87)	<0.001	1.93 (1.18–3.18)	0.009
FTD–TPI+BEV (vs. FTD–TPI)	0.56 (0.37–0.83)	0.004	0.60 (0.39–0.94)	0.025	0.74 (0.48–1.15)	0.189	1.09 (0.66–1.78)	0.742

PFS, progression-free survival; OS, overall survival; HR, hazard ratio; ECOG PS, Eastern Cooperative Oncology Group performance status; CEA, carcinoembryonic antigen; FTD–TPI+BEV, trifluridine–tipiracil plus bevacizumab; FTD–TPI, trifluridine–tipiracil.

3.4. Subgroup Analysis for Survival Outcomes

Subgroup analyses of PFS and OS comparing the efficacy of FTD–TPI+BEV and FTD–TPI are presented in Figures 2 and 3. The clinical benefits of FTD–TPI+BEV were observed across most subgroups, including those with poor prognostic factors. Patients with *RAS*-mutant disease (HR = 0.45, 95% CI, 0.27–0.74; $p = 0.001$) exhibited a more favorable response to FTD–TPI+BEV compared to those with *RAS* wild-type disease (HR = 0.85, 95% CI, 0.43–1.70; $p = 0.647$). Similarly, the treatment effect was more evident in patients with ≥3 organ metastatic sites (HR = 0.36, 95% CI, 0.21–0.63; $p < 0.001$), whereas those with ≤2 metastatic sites showed minimal benefit (HR = 0.91, 95% CI, 0.49–1.67; $p = 0.747$). Additionally, the addition of bevacizumab to FTD–TPI demonstrated clinical benefits regardless of whether the bevacizumab-free interval was shorter or longer than six months. For OS, although the differences were not statistically significant, FTD–TPI+BEV showed a trend toward improved outcomes compared to FTD–TPI alone in most subgroups. However, in patients with poor performance status, FTD–TPI+BEV was associated with worse OS outcomes (HR = 1.75, 95% CI, 0.68–4.49; $p = 0.157$). Furthermore, OS outcomes were comparable between the two treatment groups in patients with ≤2 organ metastatic sites (HR = 0.89, 95% CI, 0.44–1.80; $p = 0.736$) and those with a bevacizumab-free interval of ≤6 months (HR = 0.93, 95% CI, 0.52–1.65; $p = 0.791$).

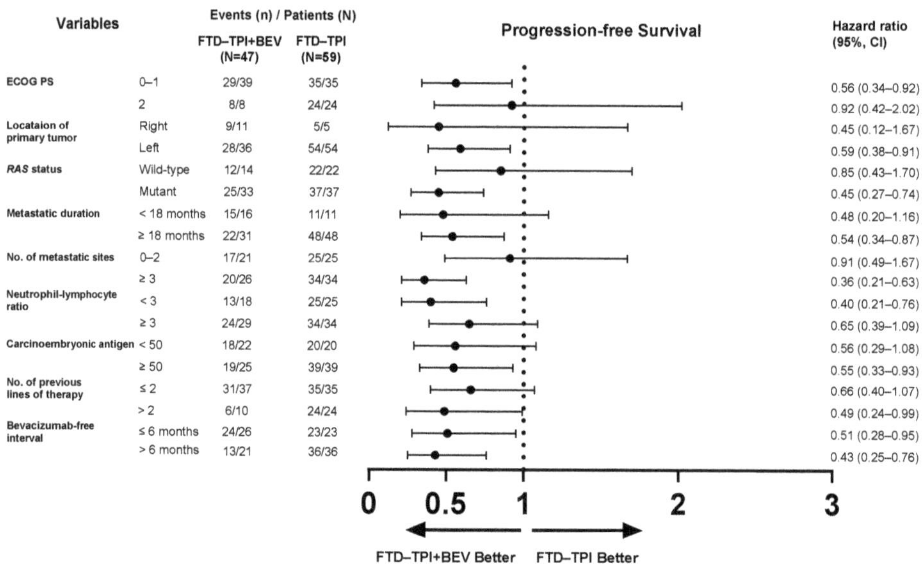

Figure 2. Forest plot of subgroup analyses for progression-free survival.

3.5. Safety

Grade 3–5 adverse events occurred in 25 patients (53.2%) in the FTD–TPI+BEV group and 32 patients (54.2%) in the FTD–TPI group (Table 4). The most frequently reported grade 3–5 adverse event was neutropenia, which affected 21 patients (38.6%) in the FTD–TPI+BEV group and 23 patients (39.0%) in the FTD–TPI group, with no notable difference between treatment groups. Febrile neutropenia was reported in two patients (4.2%) in the FTD–TPI+BEV group and two patients (3.4%) in the FTD–TPI group. Grade 3 or higher anemia was more frequently observed in the FTD–TPI group (18.6%) than in the FTD–TPI+BEV group (8.4%). Grade 1–2 nausea and fatigue occurred more frequently in the FTD–TPI+BEV group, with nausea reported in 14 patients (29.8%) compared to 8 patients (13.6%) in the FTD–TPI group and fatigue in 11 patients (23.4%) compared to 10 patients (16.9%).

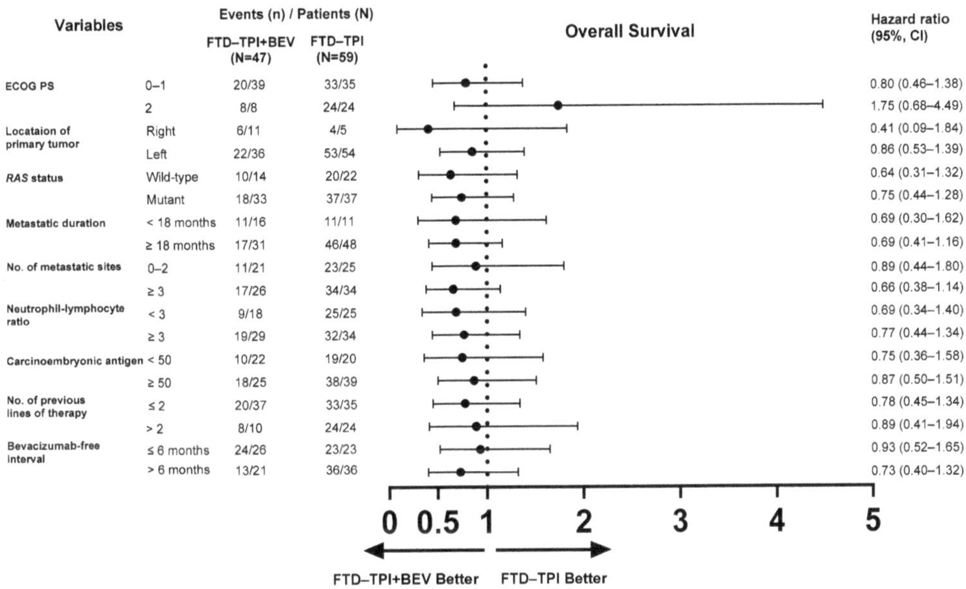

Figure 3. Forest plot of subgroup analyses for overall survival.

Table 4. Adverse events.

	FTD–TPI+BEV (*n* = 47) Any Grade	Grade ≥ 3	**FTD–TPI (*n* = 59)** Any Grade	Grade ≥ 3
Any	43 (91.5)	25 (53.2)	49 (83.0)	32 (54.2)
Diarrhea	0	0	5 (8.5)	2 (3.4)
Nausea	14 (29.8)	2 (4.2)	8 (13.6)	1 (1.7)
Vomiting	6 (12.8)	1 (2.1)	7 (11.9)	1 (1.7)
Fatigue	11 (23.4)	1 (2.1)	10 (16.9)	1 (1.7)
Neutropenia	28 (59.6)	21 (38.6)	30 (50.8)	23 (39.0)
Febrile neutropenia	2 (4.2)	2 (4.2)	2 (3.4)	2 (3.4)
Anemia	25 (53.2)	4 (8.4)	34 (57.6)	11 (18.6)
Thrombocytopenia	9 (19.1)	2 (4.2)	14 (23.7)	4 (6.8)

FTD–TPI+BEV, trifluridine–tipiracil plus bevacizumab; FTD–TPI, trifluridine–tipiracil. Data are number of patients (%).

4. Discussion

This retrospective study compared the overall effectiveness and safety of FTD–TPI+BEV and FTD–TPI monotherapy in a real-world setting for patients with mCRC refractory to standard chemotherapy. FTD–TPI+BEV demonstrated a significantly longer PFS compared to FTD–TPI monotherapy, along with a significantly higher disease control rate. However, although the FTD–TPI+BEV group exhibited a trend toward longer OS than the FTD–TPI monotherapy group, the difference was not statistically significant. Regardless of whether patients were treated with FTD–TPI+BEV or FTD–TPI alone, poor performance status, higher baseline NLR, and elevated CEA levels were significantly associated with worse OS outcomes. Additionally, FTD–TPI+BEV showed improved effectiveness outcomes across most subgroups compared to FTD–TPI monotherapy, with no significantly higher incidence of adverse events observed in the combination therapy group.

The overall study population had a similar proportion of *RAS*-mutated patients (60–70%) compared to previous trials evaluating FTD–TPI+BEV and FTD–TPI monotherapy, including the SUNLIGHT trial and a phase II trial [7,11]. However, as this was a real-world study, it included a higher proportion of patients with poor performance status (ECOG 2, 30.2%) and more heavily pretreated patients, with 40.7% having received ≥3 prior lines of therapy for metastatic disease. Furthermore, almost all patients (97.2%) had previously received anti-VEGF therapy, a markedly higher proportion than in prior trials, representing a key distinction from those studies.

The addition of bevacizumab to FTD–TPI provided a clinically meaningful advantage over FTD–TPI monotherapy in disease control and survival outcomes. However, the difference in OS was not statistically significant, which may be attributed to the lower incidence of death events in the FTD–TPI+BEV group, resulting in a substantial proportion of patients being censored (event of death: FTD–TPI+BEV vs. FTD–TPI, 54.8% vs. 100%). Given these findings, extended follow-up may provide further maturation of OS data, potentially revealing a statistically significant difference between the two groups. In patients with mCRC refractory to standard chemotherapy, prognostic factors, such as CEA, which reflects tumor burden and performance status, appeared to have a greater influence on survival outcomes. Consistent with findings from the RECOURSE trial [12], our study also identified a high NLR as an independent prognostic factor associated with inferior PFS and OS. Furthermore, the presence of peritoneal metastases was significantly correlated with worse PFS outcomes, suggesting that the effectiveness of oral chemotherapy may be limited in this subgroup of patients.

In the subgroup analysis, FTD–TPI+BEV demonstrated superior clinical efficacy compared to FTD–TPI monotherapy, regardless of primary tumor location, *RAS* mutation status, duration of metastatic disease, or number of prior lines of therapy. Notably, the survival benefit of the combination therapy was observed irrespective of the length of the bevacizumab-free interval. Consistent with prior evidence demonstrating that the continuation of bevacizumab in second-line chemotherapy improves survival outcomes after progression on first-line doublet plus bevacizumab regimens [13], our findings suggest that FTD–TPI+BEV may still offer clinical benefits even in patients who were refractory to a bevacizumab-containing regimen in their most recent treatment. However, the advantage of FTD–TPI+BEV appeared to be less pronounced in patients with poor performance status or lower tumor burden, such as those with fewer metastatic organ sites.

From a safety perspective, our findings differed from those of the SUNLIGHT trial. The addition of bevacizumab to FTD–TPI was not associated with a higher incidence of severe neutropenia, and no other clinically meaningful differences in adverse events were observed between the treatment groups. Notably, grade 3 or higher anemia was more frequently observed in the FTD–TPI monotherapy group, which may be attributable to the inclusion of patients for whom bevacizumab was contraindicated, such as those with comorbid conditions like gastrointestinal bleeding. Non-hematologic toxicities of any grade, including nausea and fatigue, were more commonly reported in the FTD–TPI+BEV group, which may be related to the longer median treatment duration in this group compared to the FTD–TPI group.

For patients with mCRC refractory to chemotherapy and without biomarkers for targeted therapy, salvage treatment options may include FTD–TPI with or without bevacizumab, regorafenib, or fruquintinib. While regorafenib, FTD–TPI, and fruquintinib have demonstrated OS benefit over placebo in previous clinical trials, FTD–TPI+BEV has shown superior OS compared to an active control, FTD–TPI monotherapy [5–7,14]. Given these findings, FTD–TPI in combination with bevacizumab may be regarded as the preferred salvage regimen in patients with good performance status. In our study, FTD–TPI+BEV demonstrated improved effectiveness outcomes regardless of the number of prior treatment lines or previous exposure to bevacizumab, supporting its use as a viable option in real-world clinical settings.

In patients for whom bevacizumab is contraindicated, such as those with a history of severe hemorrhage, impaired wound healing, or fistula formation, alternative options, including FTD–TPI monotherapy, regorafenib, or fruquintinib may be considered. Treatment selection should be made in careful consideration of the patient's clinical condition and the toxicity profile of each agent. FTD–TPI monotherapy, as observed in our real-world study, was associated with considerable hematological toxicities and should be used with caution in patients with a history of recurrent systemic infections. In contrast to FTD–TPI, regorafenib has limited hematologic toxicity but is associated with other adverse events, such as hand–foot skin reaction, proteinuria, and hypertension [6]. Nearly half of patients experience hand–foot skin reactions, underscoring the need to evaluate individual tolerability before initiating treatment [15]. Fruquintinib demonstrated a survival benefit and manageable toxicity in the FRESCO-2 trial in patients with mCRC who had progressed on or were intolerant to FTD–TPI or regorafenib [14]. Given its relatively favorable safety profile, fruquintinib may be a suitable alternative for patients with a lower disease burden and an indolent disease course who are unable to tolerate FTD–TPI or regorafenib.

Our study has several limitations. First, because the combination of bevacizumab with FTD–TPI was introduced as a treatment option following the SUNLIGHT trial results, the follow-up duration for patients treated with FTD–TPI plus bevacizumab was relatively short. Consequently, fewer death events occurred in this group, potentially limiting our ability to detect statistically significant differences in overall survival between treatment arms. Second, the relatively small patient cohort and the multiple statistical comparisons conducted increased the risk of type I errors; thus, statistically significant findings, especially those approaching significance thresholds, require careful interpretation. Moreover, the limited sample size reduced the statistical power of subgroup analyses, resulting in wide confidence intervals and decreased reliability. Additionally, it precluded advanced statistical methods, such as propensity score matching. Third, some patients experienced rapid clinical deterioration, precluding comprehensive radiologic assessment of treatment response and potentially confounding the evaluation of treatment effectiveness. Future studies with larger patient cohorts allowing for thorough radiologic assessments and employing rigorous statistical methods will be necessary to validate our findings and clarify treatment benefits.

In conclusion, our study demonstrated the effectiveness and safety of FTD–TPI with or without bevacizumab in a real-world setting among patients with chemorefractory mCRC. FTD–TPI+BEV was associated with improved survival outcomes compared to FTD–TPI monotherapy across most subgroups. Considering the varying prognostic factors, FTD–TPI+BEV may be a suitable option, particularly for selected patients with good performance status who are medically fit. Further prospective studies incorporating molecular biomarker analyses are required to better define subgroups most likely to benefit from FTD–TPI+BEV and to refine treatment sequencing strategies, including other oral chemotherapies, within the continuum of care.

Supplementary Materials: The following supporting information can be downloaded at https://www.mdpi.com/article/10.3390/biomedicines13040976/s1, Figure S1: Survival outcomes of patients with metastatic colorectal cancer treated with trifluridine–tipiracil plus bevacizumab or trifluridine–tipiracil alone as salvage therapy. (A) Progression-free survival and (B) overall survival in the entire cohort.

Author Contributions: All authors helped perform the research. S.J.P., H.K. and H.J.A. were involved in manuscript writing, draft conception and design, acquisition of data, performing procedures, and data analysis; J.H.B., Y.S.L. and I.K.L. were involved in the acquisition of data, performing procedures, and data analysis; K.S., I.-H.K. and M.L. contributed to writing the manuscript. All authors have read and agreed to the published version of the manuscript.

Funding: This research was funded by the Catholic Medical Center Research Foundation in the program year of 2024.

Institutional Review Board Statement: This study conformed to the Korean regulations and the Declaration of Helsinki. Ethical approval for the acquisition of data was obtained from the Institutional Review Board (IRB) of The Catholic University of Korea, Seoul St. Mary's Hospital (approval ID: KC25RISI0178).

Informed Consent Statement: Informed consent was waived due to the retrospective nature of the analysis.

Data Availability Statement: All materials (data and images) reported in this article are available within the paper and its Supplementary Materials.

Conflicts of Interest: The authors declare no conflicts of interest.

References

1. Bray, F.; Ferlay, J.; Soerjomataram, I.; Siegel, R.L.; Torre, L.A.; Jemal, A. Global cancer statistics 2018: GLOBOCAN estimates of incidence and mortality worldwide for 36 cancers in 185 countries. *CA Cancer J. Clin.* **2018**, *68*, 394–424. [CrossRef] [PubMed]
2. Cervantes, A.; Adam, R.; Roselló, S.; Arnold, D.; Normanno, N.; Taïeb, J.; Seligmann, J.; De Baere, T.; Osterlund, P.; Yoshino, T. Metastatic colorectal cancer: ESMO Clinical Practice Guideline for diagnosis, treatment and follow-up. *Ann. Oncol.* **2023**, *34*, 10–32. [CrossRef] [PubMed]
3. Benson, A.B.; Venook, A.P.; Adam, M.; Chang, G.; Chen, Y.-J.; Ciombor, K.K.; Cohen, S.A.; Cooper, H.S.; Deming, D.; Garrido-Laguna, I. Colon cancer, version 3.2024, NCCN clinical practice guidelines in oncology. *J. Natl. Compr. Cancer Netw.* **2024**, *22*, e240029. [CrossRef] [PubMed]
4. Pathak, P.S.; Chan, G.; Deming, D.A.; Chee, C.E. State-of-the-art management of colorectal cancer: Treatment advances and innovation. *Am. Soc. Clin. Oncol. Educ. Book* **2024**, *44*, e438466. [CrossRef] [PubMed]
5. Mayer, R.J.; Van Cutsem, E.; Falcone, A.; Yoshino, T.; Garcia-Carbonero, R.; Mizunuma, N.; Yamazaki, K.; Shimada, Y.; Tabernero, J.; Komatsu, Y. Randomized trial of TAS-102 for refractory metastatic colorectal cancer. *N. Engl. J. Med.* **2015**, *372*, 1909–1919. [CrossRef] [PubMed]
6. Grothey, A.; Van Cutsem, E.; Sobrero, A.; Siena, S.; Falcone, A.; Ychou, M.; Humblet, Y.; Bouché, O.; Mineur, L.; Barone, C. Regorafenib monotherapy for previously treated metastatic colorectal cancer (CORRECT): An international, multicentre, randomised, placebo-controlled, phase 3 trial. *Lancet* **2013**, *381*, 303–312. [CrossRef] [PubMed]
7. Prager, G.W.; Taieb, J.; Fakih, M.; Ciardiello, F.; Van Cutsem, E.; Elez, E.; Cruz, F.M.; Wyrwicz, L.; Stroyakovskiy, D.; Pápai, Z. Trifluridine–tipiracil and bevacizumab in refractory metastatic colorectal cancer. *N. Engl. J. Med.* **2023**, *388*, 1657–1667. [CrossRef] [PubMed]
8. Kagawa, Y.; Shinozaki, E.; Okude, R.; Tone, T.; Kunitomi, Y.; Nakashima, M. Real-world evidence of trifluridine/tipiracil plus bevacizumab in metastatic colorectal cancer using an administrative claims database in Japan. *ESMO Open* **2023**, *8*, 101614. [CrossRef] [PubMed]
9. Yoshino, T.; Taieb, J.; Kuboki, Y.; Pfeiffer, P.; Kumar, A.; Hochster, H.S. Trifluridine/tipiracil with or without bevacizumab in metastatic colorectal cancer: Results of a systematic review and meta-analysis. *Ther. Adv. Med. Oncol.* **2023**, *15*, 17588359221146137. [CrossRef] [PubMed]
10. Aquino de Moraes, F.C.; Dantas Leite Pessôa, F.D.; Duarte de Castro Ribeiro, C.H.; Rodrigues Fernandes, M.; Rodríguez Burbano, R.M.; Carneiro dos Santos, N.P. Trifluridine–tipiracil plus bevacizumab versus trifluridine–tipiracil monotherapy for chemorefractory metastatic colorectal cancer: A systematic review and meta-analysis. *BMC Cancer* **2024**, *24*, 674. [CrossRef] [PubMed]
11. Pfeiffer, P.; Yilmaz, M.; Möller, S.; Zitnjak, D.; Krogh, M.; Petersen, L.N.; Poulsen, L.Ø.; Winther, S.B.; Thomsen, K.G.; Qvortrup, C. TAS-102 with or without bevacizumab in patients with chemorefractory metastatic colorectal cancer: An investigator-initiated, open-label, randomised, phase 2 trial. *Lancet Oncol.* **2020**, *21*, 412–420. [CrossRef] [PubMed]

12. Argiles, G.; Yoshino, T.; Ohtsu, A.; Mayer, R.J.; Winkler, R.; Amellal, N.; Fougeray, R.; Kanehisa, A.; Van Cutsem, E. Prognostic value of neutrophil-to-lymphocyte ratio (NLR) on overall survival (OS), progression free survival (PFS) and disease control rate (DCR) in patients with metastatic colorectal cancer (mCRC) from the RECOURSE study. *J. Clin. Oncol.* **2018**, *36*, 744. [CrossRef]
13. Masi, G.; Salvatore, L.; Boni, L.; Loupakis, F.; Cremolini, C.; Fornaro, L.; Schirripa, M.; Cupini, S.; Barbara, C.; Safina, V. Continuation or reintroduction of bevacizumab beyond progression to first-line therapy in metastatic colorectal cancer: Final results of the randomized BEBYP trial. *Ann. Oncol.* **2015**, *26*, 724–730. [CrossRef] [PubMed]
14. Dasari, A.; Lonardi, S.; Garcia-Carbonero, R.; Elez, E.; Yoshino, T.; Sobrero, A.; Yao, J.; García-Alfonso, P.; Kocsis, J.; Gracian, A.C. Fruquintinib versus placebo in patients with refractory metastatic colorectal cancer (FRESCO-2): An international, multicentre, randomised, double-blind, phase 3 study. *Lancet* **2023**, *402*, 41–53. [CrossRef] [PubMed]
15. Belum, V.R.; Wu, S.; Lacouture, M.E. Risk of hand-foot skin reaction with the novel multikinase inhibitor regorafenib: A meta-analysis. *Investig. New Drugs* **2013**, *31*, 1078–1086. [CrossRef] [PubMed]

biomedicines

MDPI

Article

Prognostic Significance of *CLDN1*, *INHBA*, and *CXCL12* in Colon Adenocarcinoma: A Multi-Omics and Single-Cell Approach

Jaehwan Cheon [1,2,†], Sang Hyun Kim [3,4,†], Jaehyung Park [1] and Tae Hoon Kim [1,4,*]

1 Department of Otorhinolaryngology-Head & Neck Surgery, Korea University College of Medicine, Anam-ro 145, Seongbuk-gu, Seoul 02841, Republic of Korea
2 Department of Biomedical Science, Korea University College of Medicine, Anam-ro 145, Seongbuk-gu, Seoul 02841, Republic of Korea
3 Department of Internal Medicine, Korea University College of Medicine, Anam-ro 145, Seongbuk-gu, Seoul 02841, Republic of Korea
4 Mucosal Immunology Institute, Korea University College of Medicine, Anam-ro 145, Seongbuk-gu, Seoul 02841, Republic of Korea
* Correspondence: doctorth@korea.ac.kr
† These authors contributed equally to this work.

Abstract: Background/Objectives: Colon adenocarcinoma (COAD), the most prevalent form of colorectal cancer, remains a leading cause of cancer-related mortality. Advances in various treatments for COAD have significantly improved treatment outcomes. However, therapeutic limitations persist, highlighting the need for personalized strategies driven by novel biomarkers. The aim was to identify key hub genes that could be potential biomarkers of COAD using comprehensive bioinformatic analyses. **Methods**: Differentially expressed genes (DEGs) and co-DEGs were identified from COAD gene expression datasets. Functional enrichment analyses, including Gene Ontology (GO) and Kyoto Encyclopedia of Genes and Genomes (KEGG) pathway analysis, were performed. Hub genes were extracted from protein–protein interaction (PPI) networks and validated epigenetically using microRNA (miRNA) and DNA methylation datasets. Their expression patterns were further examined via single-cell RNA sequencing (scRNA-seq) and immune cell infiltration analysis. Prognostic relevance was assessed based on tumor metastasis and survival outcomes. **Results**: Gene expression profiling identified 118 co-DEGs, with GO and KEGG pathway analyses revealing significant pathway enrichment. PPI network analysis pinpointed 27 key co-DEGs. Epigenetic profiling indicated that both miRNA interference and DNA methylation regulate *CLDN1*, *INHBA*, and *CXCL12* expression levels. scRNA-seq analysis showed elevated *CLDN1* expression in epithelial cells and *INHBA* in myeloid cells, and reduced *CXCL12* expression in stromal cells. Prognostic analysis further demonstrated that *CLDN1* and *INHBA* are significantly associated with poor COAD outcomes. **Conclusions**: We identified some potential prognostic biomarkers for patients with COAD. Further experimental validation is required to translate these findings into precision medicine for COAD.

Keywords: colon adenocarcinoma; bioinformatics; biomarker; *CLDN1*; *INHBA*; *CXCL12*

1. Introduction

Colorectal cancer (CRC) is the third most prevalent cancer, impacting approximately 1 in every 23 men and 1 in every 25 women [1]. CRC represents 8% of all cancer-related fatalities, ranking it as the second leading cause of cancer mortality globally [2]. Significant improvements in colon cancer treatment have emerged over the years, resulting in better survival rates and quality of life for patients. The notable improvement of colon cancer treatment can be attributed to advancements in a wide array of medical treatments, including laparoscopic surgery, radiotherapy, neoadjuvant and palliative chemotherapies, and targeted therapies [3]. This progress is further supported by an enhanced understanding of the epidemiology, pathology, and molecular mechanisms associated with CRC [4]. However, there are still patients with CRC who face limitations in treatment, prompting new attempts to improve cure rates [5]. It is widely recognized that patients with CRC often exhibit varying treatment responses and prognoses, even when their tumors are histologically identical [6,7]. Consequently, personalized treatments guided by novel biomarkers are expected to yield substantial clinical efficacy and hold significant public health value [8,9].

The tumor micro-environment (TME) plays a crucial role in the progression and treatment of cancer [10,11]. TME comprises cancer cells, immune cells, stromal cells, extracellular matrix components, and signaling molecules that interact dynamically to influence tumor growth, invasion, and metastasis [11]. Targeting the TME has emerged as a promising strategy to enhance cancer treatment efficacy [12,13]. MicroRNAs (miRNAs) and DNA methylation play critical roles in the regulation of the TME [14,15]. miRNAs are short non-coding RNAs of approximately 18–25 nucleotides in length [14]. Extensive research has demonstrated the aberrant expression of miRNAs in CRC [10]. MiRNAs can function as either tumor suppressors or oncogenes in tumor tissues in CRC. In CRC, miRNAs play a crucial role in regulating and suppressing various signaling pathways, offering significant promise for diagnosis, prognosis, and personalized targeted treatment [16]. DNA methylation is an epigenetic mechanism that often leads to gene silencing when occurring in promoter regions of genes [15]. In CRC, certain crucial tumor suppressor genes can be silenced through hyper-methylation, and oncogenes can be activated by hypomethylation processes [15].

As bioinformatics techniques become more advanced and sophisticated, the discovery and characterization of differentially expressed genes (DEGs) as hub genes in diseases like colon adenocarcinoma (COAD)—which constitutes around 95% of CRC—are accelerating. Although several previous studies have explored biomarkers in COAD using transcriptomic or epigenomic data independently, few have attempted to comprehensively integrate multi-omics data—including gene expression, miRNA regulation, DNA methylation, and single-cell RNA sequencing (scRNA-seq)—to identify robust and clinically relevant hub genes. This study is among the first to combine these diverse analytical layers with immune infiltration and prognostic analyses to systematically investigate the TME and molecular mechanisms of COAD.

In this study, we have focused on hub genes that interact with miRNAs and methylation changes, identifying novel biomarkers that are not only diagnostic but also predictive of response to immunotherapies, ultimately advancing personalized medicine in COAD oncology. Using five GSE (gene expression omnibus series) datasets composed of colon tissue, we identified hub genes of COAD. We then utilized the COAD miRNA dataset to identify significant miRNAs and correlated them with the previously identified hub genes. We also analyzed the methylation status of hub genes through bioinformatics, classifying genes with significant methylation changes. We conducted further immunologic analysis on the selected hub genes to confirm their potential as valuable biomarkers for future use.

2. Materials and Methods

2.1. Microarray Data

The Gene Expression Omnibus (GEO), hosted by the National Center for Biotechnology Information (NCBI), is an openly accessible repository found at https://www.ncbi.nlm.nih.gov/geo/ (accessed on 15 January 2024). We retrieved five COAD gene expression datasets from GEO using keywords such as 'Colorectal cancer' and 'COAD' and analyzed them via GEO2R (https://www.ncbi.nlm.nih.gov/geo/info/geo2r.html, accessed on 15 January 2024) [17]. These datasets include GSE37364 (10 COAD tissues and 10 normal colonic mucosa), GSE41657 (25 COAD tissues and 12 normal epithelium or colorectal mucosa), GSE44076 (98 COAD tissues and 50 normal mucosa), GSE110224 (17 COAD tissues and 17 normal tissues), and GSE115261 (10 COAD tissues and 10 normal tissues). Since datasets used were obtained from publicly available repository, ethical review and approval were waived for this study.

2.2. Identification of Differentially Expressed Genes (DEGs) and Co-DEGs

Using GEO2R, we analyzed DEGs between COAD and normal tissues across all five datasets. We identified DEGs within each dataset (GSE37364, GSE41657, GSE44076, GSE110224, and GSE115261) based on an adjusted p-value < 0.05, and a $\log_2$ fold change ($\log_2$FC) threshold of $|\log_2 FC| > 1.5$. After detecting DEGs in each dataset, a cross-analysis of these five datasets was performed using a Venn diagram to detect co-DEGs.

2.3. Gene Ontology (GO) and Kyoto Encyclopedia of Genes and Genomes (KEGG) Pathway Analysis of Up- and Down-Regulated Co-DEGs

GO and KEGG enrichment analyses were conducted using the Database for Annotation, Visualization, and Integrated Discovery (DAVID) (https://david.ncifcrf.gov, accessed on 5 February 2024) (v7.0) [18].

2.4. Protein–Protein Interaction (PPI) Network Buildup on Up- and Down-Regulated DEGs for Hub Genes Detection

PPI networks were constructed using STRING (v12.0) (https://string-db.org/, accessed on 5 February 2024) [19].

2.5. Analysis of Differentially Expressed miRNAs (DEMs) Related to the Hub Gene Expression

We detected DEMs using the criteria of an adjusted p-value < 0.05 and threshold of $|\log_2 FC| > 1$. These analyses were conducted across three selected non-coding RNA profiling datasets: GSE18392 (116 colon tumors and 29 normal colon samples), GSE35982 (8 colorectal cancer tumors and 8 normal colorectal tissue samples), and GSE41655 (33 COAD tissues and 15 normal colorectal mucosa samples) obtained from GEO2R. Following the identification of up-regulated and down-regulated DEMs and co-DEMs across three datasets, we utilized The University of Alabama at Birmingham CANcer data analysis Portal (UALCAN) database (https://ualcan.path.uab.edu/analysis.html, accessed on 19 February 2024), which operates based on The Cancer Genome Atlas (TCGA) [20] for validating the expression levels of the co-DEMs in COAD and identification of key DEMs. Subsequently, we investigated the target genes of key co-DEMs to assess their influence on hub gene expression in COAD using ENCORI/starBase (v3.0) (https://rnasysu.com/encori/, accessed on 19 February 2024) [21]. As miRNA typically inhibits the transcription of target genes [22], we examined the negative correlation between expression of these key co-DEMs and hub genes.

2.6. Analysis of Differentially Methylated Regions (DMRs) of Hub Genes

We identified DMRs of the CpG island, expression-regulatory elements of a gene, such as promoters and enhancers, using the criteria of an adjusted p-value < 0.05 and a threshold of $|\log_2FC| > 0.015$. This analysis was conducted on the GSE42752 dataset, which includes 22 COAD and 41 normal colon genomic DNA samples obtained from GEO2R. As hyper-methylated DMRs of specific genes typically lead to the transcription inhibition of the gene [23], we analyzed the negative correlation between methylation levels of DMRs and expression levels of hub genes. Moreover, we corroborated the level of promoter methylation and expression of hub genes by scrutinizing UALCAN for using TCGA dataset and validation.

2.7. Hub Gene Expression Pattern Analysis in the COAD scRNA-Seq Dataset

For exploring the expression patterns based on cell types of some hub genes in COAD, we opted for the GSE178341 dataset, consisting of 62 COAD tumor (total 258,359 cells) and 36 normal colon tissues (total 112,864 cells) using the Single-Cell Portal (SCP) database (https://singlecell.broadinstitute.org/single_cell, accessed on 12 February 2024) [24].

2.8. Analysis of Immune Cell Infiltration Level

We utilized the TIMER 2.0 database (http://timer.cistrome.org/, accessed on 12 February 2024) to compute the correlations between the expression levels of hub genes and the infiltration of various immune cell types in COAD (n = 458) [25].

2.9. Tumor Metastasis Analysis According to Hub Gene Expression

We assessed the influence of specific hub gene expression on tumor metastasis in patients with COAD to investigate its effect on tumor prognosis using the Tumor, Normal, and Metastatic tissues (TNM) plot.com database (377 normal, 1450 tumor, and 99 metastatic samples) (https://tnmplot.com/, accessed on 19 February 2024) [26].

2.10. Analysis of Survival Rates According to Hub Gene Expression

We employed the Kaplan–Meier (KM) plot database (https://kmplot.com/, accessed on 19 February 2024) to explore the correlation between the expression levels of specific hub genes and patient survival in individuals with COAD [27]. It was utilized to assess overall survival (OS) in patients with COAD (n = 1061) for certain hub genes.

2.11. Data Visualization

All volcano plots, heatmap plots, and bar and bubble plots, illustrating the GO and KEGG pathways, were generated using Hiplot (https://hiplot-academic.com/, accessed on 8 February 2024) [28].

Taken together, Figure 1 exhibits the overall research flow used in this study.

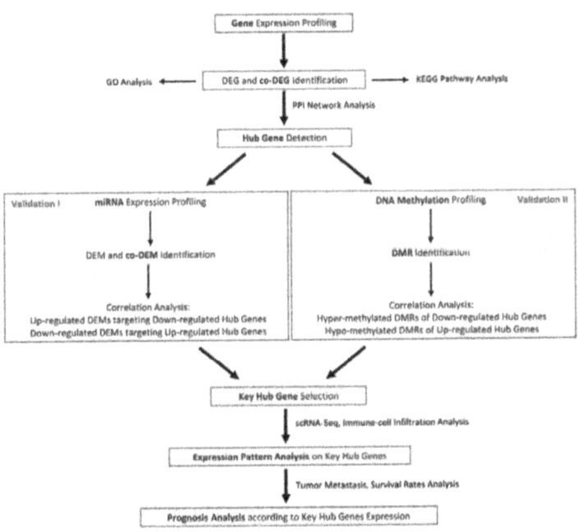

Figure 1. Flowchart used in this study.

3. Results

3.1. Identification of DEGs and Co-DEGs in Five COAD Gene Datasets

Using GEO2R, we analyzed five datasets (GSE37364, GSE41657, and GSE44076, GSE110224, GSE115261) and identified a total of 3329 up-regulated and 4782 down-regulated DEGs. Reciprocal volcano maps for each dataset illustrate the distribution of significantly altered genes (Figure 2A). Representative heatmaps showcase 20 DEGs in each dataset (Figure 2A). Notably, cross-analysis revealed 118 co-DEGs (38 up-regulated and 80 down-regulated), visualized in a Venn diagram (Figure 2B).

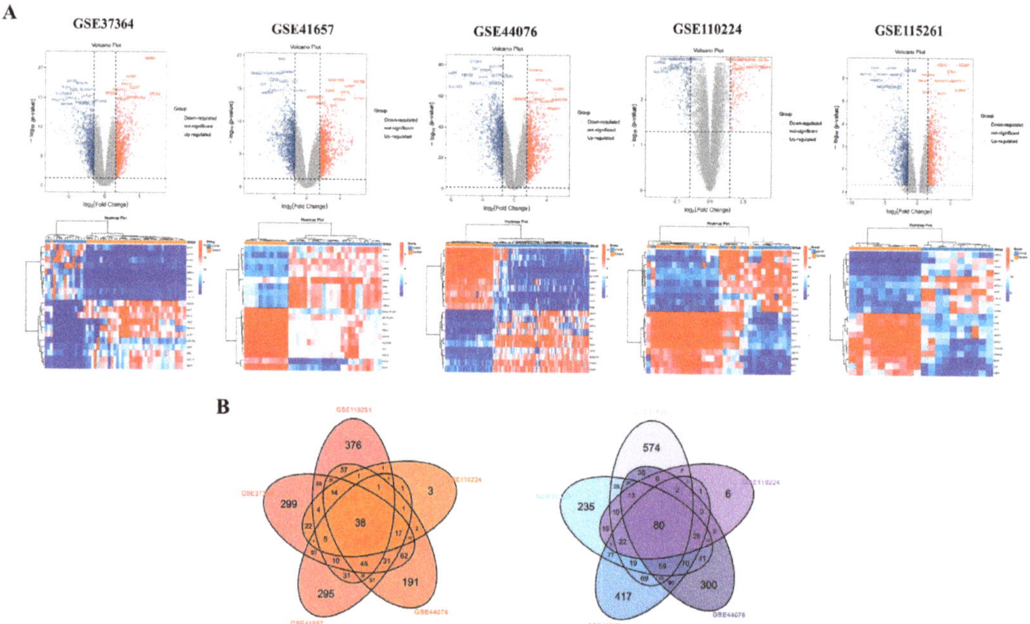

Figure 2. Identification of DEGs and detection of co-DEGs of COAD using five gene expression datasets. Volcano map illustrating differentially expressed gene (DEG) distribution and heatmap showing representative 10 up-regulated and down-regulated DEGs in GSE37364, GSE41657, GSE44076, GSE110224, and GSE115261 datasets (**A**). The red points in the volcano plots indicate up-regulated genes identified with a fold change of $\geq$1.5 and a corrected p-value of <0.05. Conversely, the blue points represent down-regulated genes identified with a fold change ≤ -1.5 and a corrected p-value < 0.05. Black points denote genes with no statistically significant difference. Gene expression is depicted in a heatmap using color coding. Red indicates up-regulation, blue denotes down-regulation, and white indicates no significant change. A total of 38 up-regulated and 80 down-regulated co-DEGs are identified through analysis of the cross-linking data from GSE37364, GSE41657, GSE44076, GSE110224, and GSE115261, and visualized using a Venn diagram (**B**).

3.2. GO and KEGG Pathway Analysis on Up- and Down-Regulated Co-DEGs

To comprehend the functional implications of co-DEGs, we conducted GO and KEGG pathway analyses for each up- and down-regulated co-DEGs. As illustrated in Figure 3A, up-regulated co-DEGs showed significant enrichment in biological pathways (BPs) related to proteolysis, extracellular matrix (ECM) organization, inflammatory response, and other associated processes. Additionally, they were notably enriched in cellular components (CCs) associated with the extracellular space, extracellular region, membrane, and other components. Moreover, they exhibited significant enrichment in molecular functions (MFs) related to identical protein binding, zinc ion binding, and serine-type endopeptidase activity, among others. KEGG pathway analysis further underscored specific pathways enriched with up-regulated co-DEGs. These included significant enrichment in rheumatoid arthritis, cytokine–cytokine receptor interaction, the interleukin (IL)-17 signaling pathway, and others (Figure 3B).

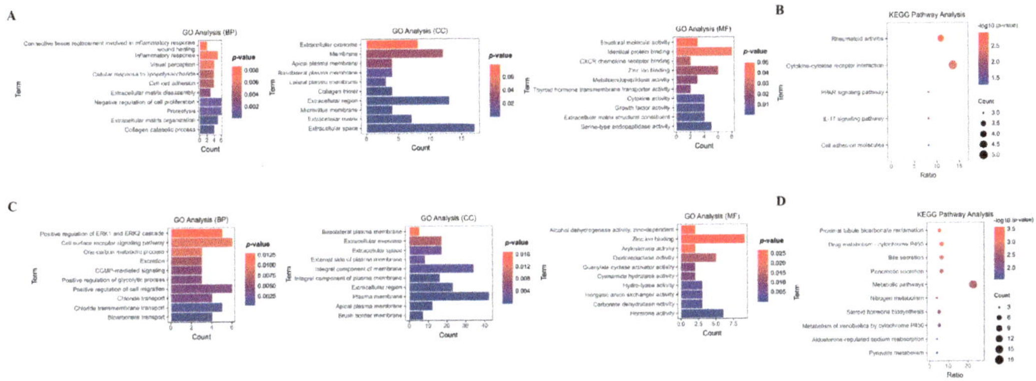

Figure 3. GO and KEGG analyses of co-DEGs in the five datasets. GO analysis categorized the up-regulated and down-regulated co-DEGs into several biological pathways (BPs), cellular components (CCs), and molecular functions (MFs), based on their roles (**A**,**B**). Furthermore, KEGG pathway analysis is used to classify the up-regulated and down-regulated co-DEGs biochemical pathways according to their gene functions (**C**,**D**).

Conversely, down-regulated co-DEGs demonstrated significant enrichment in biological pathways associated with positive regulation of cell migration, cell surface receptor signaling pathways, positive regulation of extracellular signal-regulated kinase (ERK) 1 and ERK2 cascades, and other related processes. Additionally, they were notably enriched in cellular components associated with the plasma membrane, integral components of the membrane, extracellular regions, and other components. Moreover, they exhibited significant enrichment in molecular functions related to zinc ion binding, hormone activity, oxidoreductase activity, and other functions (Figure 3C). KEGG pathway analysis revealed enrichment in metabolic pathways, steroid hormone biosynthesis, bile secretion, and other pathways (Figure 3D).

3.3. PPI Network Construction of Co-DEGs and Detection of Hub Genes

For the identification of key genes potentially influencing the progression of COAD, we scrutinized all 118 co-DEGs using the STRING database to construct PPI networks. Co-DEGs showing connectivity above six were considered hub genes, unveiling numerous promising contenders. Notably, insulin-like growth factor (IGF)1 emerged with the highest connectivity at sixteen, followed by matrix metalloproteinase (MMP)1 at fourteen, and others, including cluster of differentiation (CD)36 (node degree of thirteen), collagen type I alpha 1 chain (COL1A1) (node degree of eleven), C-X-C motif chemokine ligand (CXCL)12 (node degree of eleven), claudin (CLDN)1 (node degree of six), and inhibin β (INHB)A (node degree of six). In total, 27 hub genes were filtered from the 118 co-DEGs (minimum required interaction score = 0.4, $p < 1.0 \times 10^{-16}$) (Figure 4A). Additionally, PPI networks of up-regulated (minimum required interaction score = 0.4, $p < 1.0 \times 10^{-16}$) (Figure 4B) and down-regulated co-DEGs (minimum required interaction score = 0.4, $p < 1.0 \times 10^{-16}$) (Figure 4C) were constructed.

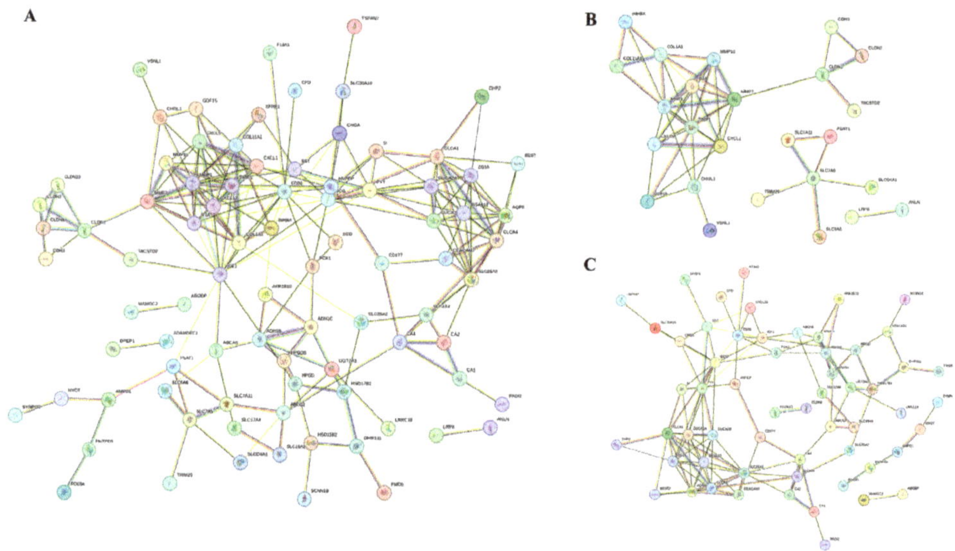

Figure 4. PPI network diagrams of co-DEGs. Each diagram represents a network of all co-DEGs (**A**), up-regulated DEGs (**B**), and down-regulated DEGs (**C**). Each network represents a gene, and each line represents the interaction of proteins. The results within the circle represent protein structure. The color of the line represents evidence of the interaction.

3.4. Identification of DEMs Regulating Hub Gene Expression in Three COAD Datasets

Using GEO2R, we analyzed three datasets (GSE18392, GSE35982, and GSE41655) and identified a total of 240 up-regulated and 144 down-regulated DEMs. Reciprocal volcano maps for each dataset illustrate the distribution of significantly altered miRNAs (Figure 5A). Notably, cross-analysis revealed 8 co-DEMs consisting of up-regulated hsa-miR-135b (also known as has-miR-135b-5p), hsa-miR-183, hsa-miR-224, and hsa-miR-552 and down-regulated hsa-miR-30a, hsa-miR-375, hsa-miR-378a (also known as hsa-miR-378*), and hsa-miR-551b visualized in a Venn diagram (Figure 5B).

Consistent with our up-regulated DEMs, the TCGA dataset also showed significantly higher expression of all four DEMs in COAD tumors compared to the control (hsa-miR-135b: $p < 1 \times 10^{-12}$, hsa-miR-183: $p < 1 \times 10^{-12}$, hsa-miR-224: $p = 2.7479 \times 10^{-9}$, and hsa-miR-552: $p = 1.6245 \times 10^{-12}$) (Figure 5C). We examined the negative correlation between these key co-DEMs and hub gene expression to investigate the impact of up-regulated co-DEMs on down-regulated hub genes. Out of the 4 up-regulated co-DEMs, only has-mir-135b-5p exhibited a negative correlation, with the expression levels of *CD36* (r = −0.238, $p = 3.38 \times 10^{-7}$) and *CXCL12* (r = −0.408, $p = 1.70 \times 10^{-19}$), which were down-regulated hub genes (Figure 5D). By contrast, analysis of the TCGA dataset indicated that the key down-regulated co-DEMs were hsa-miR-375 ($p < 1 \times 10^{-12}$) and hsa-miR-378 ($p = 1.61614 \times 10^{-5}$), as these DEMs exhibited significant decreases in tumor compared to the control, while no significant difference was observed for hsa-miR-30a ($p = 6.7714 \times 10^{-1}$) and hsa-miR-551b ($p = 3.981 \times 10^{-1}$) between the groups (Figure 5E). To assess the influence of key down-regulated co-DEMs on up-regulated hub genes, we examined the negative correlation between these DEMs and their target genes. Among two key down-regulated DEMs, only hsa-miR-375 displayed a negative correlation, with the expression levels of *CLDN1* (r = −0.278, $p = 1.97 \times 10^{-9}$) and *INHBA* (r = −0.223, $p = 1.86 \times 10^{-6}$), which were up-regulated hub genes (Figure 5F).

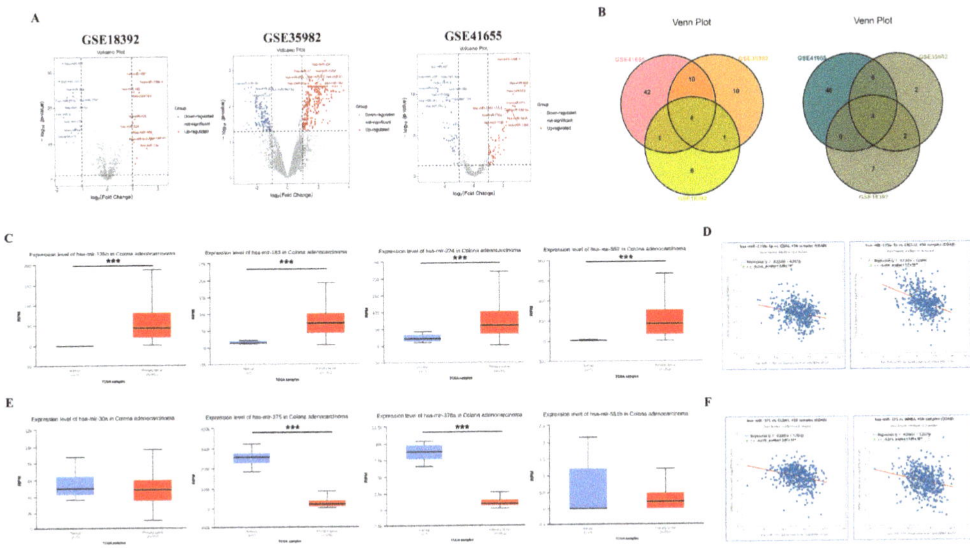

Figure 5. Some hub gene expression levels are regulated by miRNA interference. Volcano maps showing the distribution of differently expressed microRNAs (miRNAs) (DEMs) in GSE18392, GSE35982, and GSE41655 (**A**). In the volcano plots, red points indicate significantly up-regulated DEMs identified with a fold change $\geq$ 1.0 and a corrected p-value of <0.05. On the contrary, blue points represent down-regulated DEMs identified with a fold change ≤ -1.0 and a corrected p-value of <0.05. The co-DEMs in the three miRNA datasets were separated using a Venn diagram of the up-regulated and down-regulated DEMs, respectively (**B**). All four up-regulated co-DEMs are expressed at a higher level in the TCGA database (**C**), and only hsa-135b-5p expression is negatively correlated with the expression of some down-regulated hub genes, including *CD36* and *CXCL12* in patients with COAD (**D**). Conversely, hsa-miR-375 and hsa-miR-378a are significantly decreased in COAD tumors compared to the controls among the four down-regulated co-DEMs in the TCGA database (**E**). Among these miRNAs, only the hsa-miR-375 expression is negatively correlated with some up-regulated hub gene expression, including *CLDN1* and *INHBA*, in patients with COAD (**F**). *** p-value < 0.001.

3.5. Identification of DMRs Modulating the Expression of Hub Genes

We conducted methylation profiling analysis on COAD tumor and normal tissue using GSE42752 dataset, revealing a total of 179,879 DMRs, comprising 78,343 hyper-methylated and 101,536 hypo-methylated CpG sites (Figure 6A). we analyzed the inverse correlation between methylation levels of DMRs and expression levels of hub genes to investigate the impact of methylation on hub gene expression. Among the hyper-methylated DMRs, cg14240353 and cg26267854 were identified in the enhancer and promoter regions of *IGF1* and *CXCL12*, respectively, which are down-regulated hub genes. Hypo-methylated DMRs, including cg14543953, cg02061229, cg27604897, cg27606396, cg15105660, and cg05885137, were located in the enhancer regions of *MMP1*, *MMP10*, *COL1A1*, *IL1A*, *CLDN1*, and *INHBA*, all of which are up-regulated hub genes (Figure 6B). In line with our findings of hyper-methylated DMRs in *CXCL12*, analysis of the TCGA dataset also revealed significant hyper-methylation of the *CXCL12* promoter ($p = 1.62425 \times 10^{-12}$) in COAD tumors compared to controls. However, there was no significant difference observed between the two groups in the case of *IGF1* promoter ($p = 9.357 \times 10^{-1}$). Conversely, analysis of the TCGA dataset revealed significant promoter hypo-methylation of *MMP1* ($p = 1.62448 \times 10^{-12}$), *MMP10* ($p = 4.6901 \times 10^{-2}$), *COL1A1* ($p = 5.0124 \times 10^{-4}$), *IL1A* ($p = 1.41809 \times 10^{-12}$), *CLDN1*

($p = 3.33067 \times 10^{-16}$), and *INHBA* ($p = 1.62448 \times 10^{-12}$), which is consistent with our findings from the DMR methylation analysis of the up-regulated hub genes (Figure 6C).

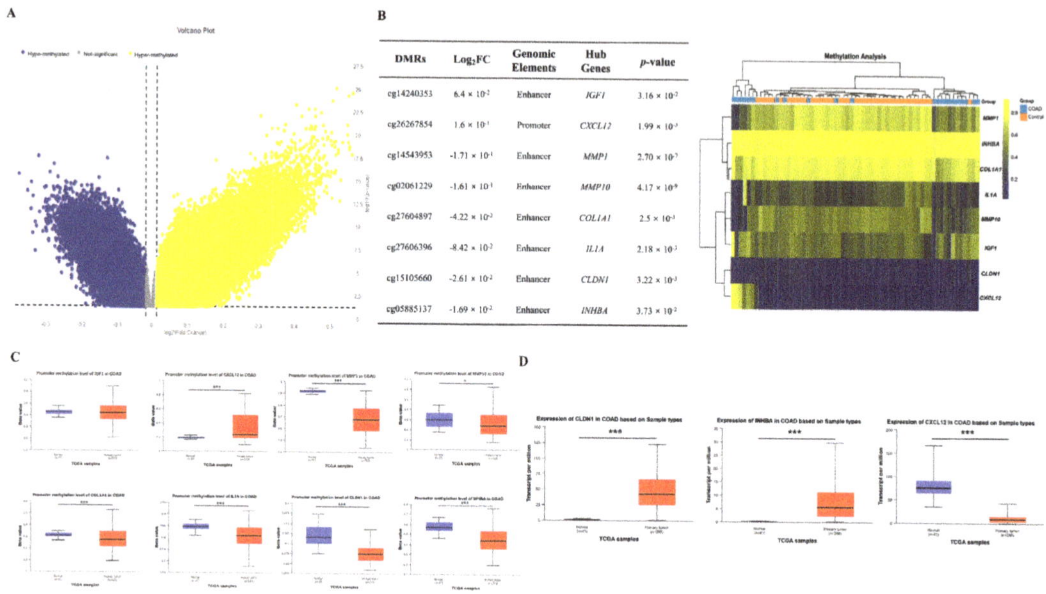

DMRs	Log_2FC	Genomic Elements	Hub Genes	*p*-value
cg14240353	6.4×10^{-2}	Enhancer	*IGF1*	3.16×10^{-2}
cg26267854	1.6×10^{-1}	Promoter	*CXCL12*	1.99×10^{-3}
cg14543953	-1.71×10^{-1}	Enhancer	*MMP1*	2.70×10^{-7}
cg02061229	-1.61×10^{-1}	Enhancer	*MMP10*	4.17×10^{-9}
cg27604897	-4.22×10^{-2}	Enhancer	*COL1A1*	2.5×10^{-3}
cg27606396	-8.42×10^{-2}	Enhancer	*IL1A*	2.18×10^{-3}
cg15105660	-2.61×10^{-2}	Enhancer	*CLDN1*	3.22×10^{-3}
cg05885137	-1.69×10^{-2}	Enhancer	*INHBA*	3.73×10^{-2}

Figure 6. Expression levels of hub genes controlled by epigenetic regulation. Volcano map illustrating differently methylated region (DMR) distribution in GSE42752. In the volcano plots, yellow dots denote notably hyper-methylated DMRs identified with a fold change of ≥ 0.015 and a corrected *p*-value of <0.05. Conversely, blue dots represent hypo-methylated DMRs identified with a fold change ≤ -0.015 and a corrected *p*-value < 0.015 (**A**). Among the DMRs of various genes, the DMRs of hub genes with a negative correlation between methylation and expression are listed, and the methylation levels of these genes are depicted in a heatmap using color coding. Yellow signifies hyper-methylation, blue denotes hypo-methylation, and gray indicates no significant change (**B**). The promoter methylation levels of all hub genes listed in (**B**) exhibited significant differences between COAD tumor and control except for the *IGF1* in the TCGA database (**C**). Additionally, the expression levels of *CLDN1*, *INHBA*, and *CXCL12* are hub genes affected by both miRNAs and DNA methylation among hub genes. Analysis of the TCGA database revealed significantly increased expression levels of *CLDN1* and *INHBA*, while *CXCL12* exhibited significantly reduced expression levels (**D**). * *p*-value < 0.05, *** *p*-value < 0.001.

Remarkably, the expression levels of *INHBA*, *CLDN1*, and *CXCL12* among hub genes exhibited a pronounced negative correlation with both miRNAs targeting these genes and the DNA CpG site methylation regulating the epigenetic features of these genes. Additionally, the TCGA database indicated significantly higher expression of *CLDN1* and *INHBA*, while *CXCL12* displayed lower expression between COAD tumor and control samples, consistent with our analysis results of the DEGs, DEMs, and DMRs. (Figure 6D).

3.6. Expression Pattern Analysis of Hub Genes Using COAD scRNA-Seq Dataset

We performed scRNA-seq analysis on *CLDN1*, *INHBA*, and *CXC12*, regulated by both miRNAs and DNA methylation, in COAD tumor and normal colon tissue using GSE178341. Figure 7A illustrates the overall cell types of patients with COAD and control. The dot plot showed that the expression of *CLDN1*, *INHBA*, and *CXCL12* exhibited identical expression regulation with our previous analysis results in overall cell types (*CLDN1* and *INHBA* were up-regulated, and *CXCL12* was down-regulated, in COAD compared to the control. *CLDN1* is primarily expressed in epithelial cells, and *INHBA* is predominantly expressed in myeloid cells, with *CLDN1* and *INHBA* displaying increased expression levels in epithelial cells and myeloid cells in COAD compared to the control, respectively (Figure 7B,C). Conversely, *CXCL12* was mainly expressed in stromal cells and exhibited decreased expression levels in these cell types (Figure 7D). Among these three hub genes, only *INHBA* showed association with immune cell types, particularly being prominently expressed in monocytes among various myeloid cell types, such as dendritic cells (DCs), granulocytes, and macrophages (Figure 7E).

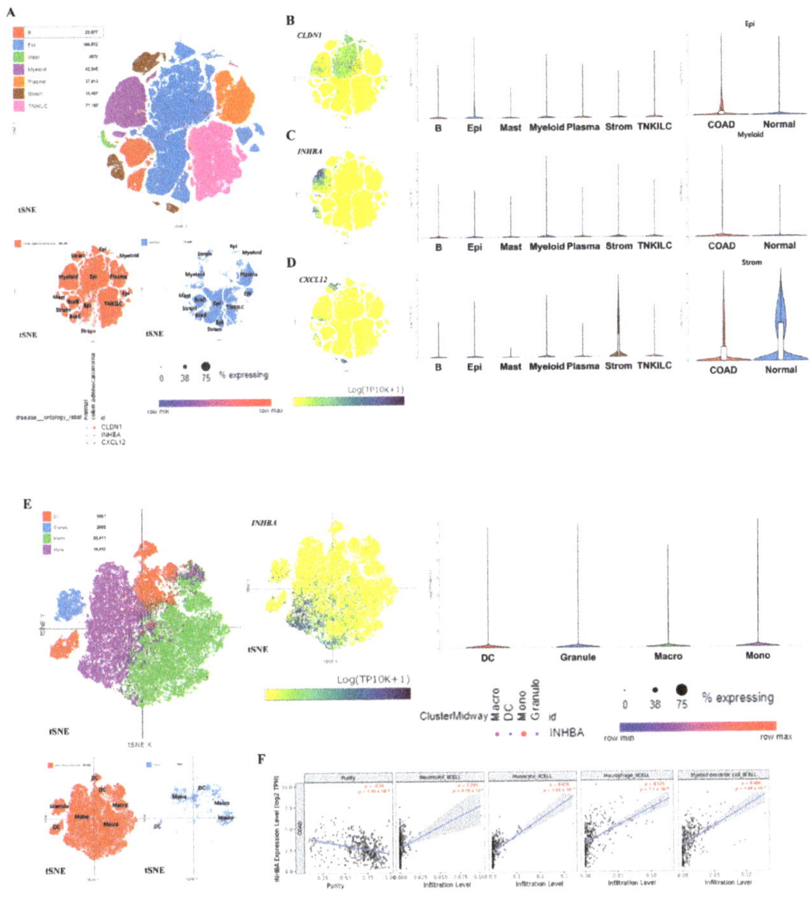

Figure 7. Expression patterns of the key hub genes and their correlation with immune cell infiltration. The entire cell types of COAD tumors and normal tissues and the expression levels of the three hub genes in all cell types are presented (**A**). Expression patterns of three hub genes and expression levels between COAD and control in the mainly expressed cell types (**B**–**D**). Expression pattern of *INHBA* in myeloid cells (**E**). All single-cell RNA sequencing data are depicted using the t-distributed Stochastic Neighbor Embedding (t-SNE) method. Immune infiltration patterns based on *INHBA* expression in myeloid cell types in COAD (**F**).

3.7. Correlation Analysis Between Immune Cell Infiltration and INHBA Expression Levels

To investigate the relevance between immune cell infiltration, particularly in cells expressing *INHBA*, and the level of *INHBA* expression, we calculated Spearman's rho values on these elements in 458 COAD samples using the XCELL algorithm (Figure 7F). Consistent with the scRNA-seq results on *INHBA* expression levels in myeloid cells, monocytes (Spearman $\rho = 0.476$, $p = 5.65 \times 10^{-17}$) exhibited a significantly positive correlation between infiltration and *INHBA* expression levels in COAD. Furthermore, other myeloid cells, such as macrophages (Spearman ρ = 0.525, $p = 7.10 \times 10^{-21}$), granulocytes (especially neutrophils) (Spearman ρ = 0.235, $p = 8.19 \times 10^{-5}$), and DCs (Spearman ρ = 0.484, $p = 1.44 \times 10^{-17}$), also showed a notably positive correlation according to gene expression.

3.8. Tumor Progress Analysis Based on the Expression of Hub Genes

Figure 8A exhibited the relationship between expression of hub genes and tumor prognosis. Both *CLDN1* ($p = 2.94 \times 10^{-133}$) and *INHBA* ($p = 5.52 \times 10^{-19}$) expressions were significantly up-regulated in a tumor progression-dependent manner. However, in the case of *CXCL12*, the expression levels did not show down-regulation in a tumor progression-dependent manner ($p = 6.88 \times 10^{-72}$). Furthermore, Figure 8B demonstrated the relationship between hub gene expression and survival rates in patients with COAD. Consistent with the tumor metastatic analysis, high expression of *CLDN1* (hazard ratio (HR) > 1.3; $p < 0.01$) and *INHBA* (HR > 1.5; $p < 0.001$) was associated with poor prognosis, as indicated by a high HR. Additionally, high expression of *CXCL12* was also linked to poor survival rates (HR > 1.4; $p < 0.001$), despite our data suggesting that *CXCL12* was a down-regulated co-DEG.

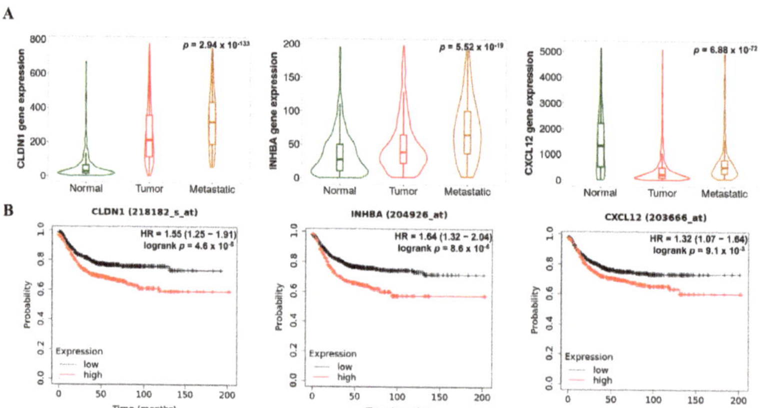

Figure 8. Prognostic patterns in COAD according to the expression levels of the key hub genes. Metastatic effects (**A**) and overall survival (OS) rates (**B**) based on *CLDN1*, *INHBA*, and *CXCL12* expression levels.

4. Discussion

Understanding the underlying molecular mechanisms of COAD in terms of the TME would greatly benefit diagnosis, management, and prognosis evaluation. TME is a dynamic and complex network surrounding a tumor, consisting of various cell types, signaling molecules, and extracellular matrix components [29,30]. This environment plays a crucial role in tumor development, progression, and response to therapy. Two key regulatory mechanisms within the TME are miRNAs and DNA methylation, both of which significantly influence gene expression and cellular behavior [31,32]. The current study integrates extensive bioinformatics analysis to uncover novel biomarkers and molecular pathways associated with CRC, with a specific focus on COAD, which comprises the majority of CRC cases. By analyzing differential gene expression, miRNA regulation, and DNA methylation patterns, we identified key hub genes and pathways that may serve as critical therapeutic targets for personalized treatments.

Our analysis suggested the possible role of DEMs in regulating key hub genes. For example, hsa-miR-135b-5p showed negative correlation with *CD36* and *CXCL12*, both of which are down-regulated in COAD. hsa-miR-375 also displayed a negative correlation with up-regulated hub genes *CLDN1* and *INHBA*. These findings are consistent with previous reports demonstrating the involvement of miRNAs in tumorigenesis through post-transcriptional regulation, impacting processes like cell migration and invasion, and immune cell infiltration [33,34].

Epigenetic changes, specifically DNA methylation, emerged as another crucial regulatory mechanism influencing the expression of hub genes. We identified multiple DMRs in CpG islands, and enhancer and promoter regions of hub genes, correlating with altered gene expression. Notably, *CXCL12* exhibited hyper-methylation in their promoter regions, contributing to their down-regulation in COAD. Conversely, *MMP1*, *CLDN1*, *COL1A1*, and *INHBA* showed hypo-methylation in their enhancer regions, aligning with their up-regulation. These findings are consistent with the well-established role of DNA methylation in cancer, where hyper-methylation of tumor suppressor genes and hypo-methylation of oncogenes drive tumorigenesis [31].

We identified three hub genes—*INHBA*, *CLDN1*, and *CXCL12*—that showcase intricate regulatory processes involving both miRNA regulation and DNA methylation. These mechanisms are crucial for their expression levels and play a significant role in influencing tumor dynamics.

INHBA, part of the transforming growth factor β superfamily, is encoded in the nucleus of human cells, synthesized in the cytoplasm, and secreted through the membrane [35]. Recent research indicates that the *INHBA* is over-expressed in various cancers and is associated with cell proliferation and outcomes in lung [36], gastric [37], esophageal [38], and colorectal tumors [39]. A study on esophageal adenocarcinoma found higher *INHBA* expression in cancerous tissues compared to hyper-plastic esophageal tissues [38]. This suggests that *INHBA* over-expression may enhance cell proliferation and be influenced by promoter demethylation and histone acetylation in esophageal adenocarcinoma cell lines. Studies showed that overexpression of *INHBA* is positively correlated with poor prognosis in esophageal, prostate, and ovarian cancer [36,40,41]. In our study, through prognostic analysis, *INHBA* was recognized as correlating with poor prognosis in CRC patients. *INHBA* is often over-expressed in COAD tissues, correlating with increased tumor aggressiveness, higher metastatic potential, and poorer prognosis [42].

hsa-miR-375 functions as a tumor suppressor in many types of cancer. In colon cancer, hsa-miR-375 is often down-regulated, and its reduced expression is linked to poor patient prognosis [43,44]. Several studies have shown that while direct effects in CRC have not been definitively identified, hsa-miR-375 may target *INHBA* and potentially act to suppress its expression [45,46]. The reduction in hsa-miR-375 levels may diminish its suppressive impact on *INHBA*, potentially causing an upsurge in *INHBA* expression in CRC. Such a dysregulation could lead to more aggressive tumor characteristics, including increased cell proliferation and invasion, and a greater propensity for metastasis. The inverse relationship between *INHBA* and hsa-miR-375 in our study suggests that restoring the levels of hsa-miR-375 could potentially suppress *INHBA* activity, offering a therapeutic approach to inhibit tumor progression in colon cancer. This interaction highlights the importance of both *INHBA* and hsa-miR-375 as potential biomarkers and targets for personalized treatment strategies in CRC.

The CLDN family comprises at least 24 members, with their expression varying according to cell type [47]. CLDN1 is a crucial element of tight junctions and is vital in tumorigenesis [48]. CLDNs are responsible for regulating the differentiation, proliferation, and migration of epithelial cells [49]. Recent studies indicate that the expression of CLDN genes is frequently altered in cancers [50,51]. The role of CLDNs in cancer remains unclear; however, recent research suggests that the CLDN1-dependent pathway might play a role in suppressing CRC expression and is associated with tumor invasiveness and prognostic factors [49]. Studies have found that CLDN1 mRNA expression is elevated in CRC compared to normal colonic mucosa [52]. Moreover, CLDN1 has been linked to colon cancer tumorigenesis [53]. However, various studies have reported that claudin levels in cancer vary, with some showing increased expression [54,55] and others showing decreased expression levels [56,57]. In our study, *CLDN1* expression was significantly up-regulated in a tumor progression-dependent manner. Higher expression of CLDN1 was significantly associated with poor outcome.

Similarly to INHBA, our analysis showed an inverse relationship between *CLDN1* and hsa-miR-375. Over-expression of hsa-miR-375 down-regulates *CLDN1*, while knockdown of hsa-miR-375 up-regulates *CLDN1* in non-small cell lung cancer [58]. It is believed that a similar mechanism may operate in colon cancer. The down-regulation of hsa-miR-375, which is often observed in COAD, can lead to the up-regulation of *CLDN1*. This up-regulation can disrupt cell–cell adhesion due to changes in tight junction composition, facilitating enhanced cancer cell migration and invasion. The increased expression of *CLDN1* in CRC has been associated with poorer prognosis and may serve as a biomarker for invasive disease characteristics. Furthermore, the restoration of hsa-miR-375 levels might represent a therapeutic approach to mitigate these effects by repressing *CLDN1* expression, potentially inhibiting tumor progression and improving patient outcomes.

CXCL12 is an α-chemokine derived from stromal cells that encodes a family of interstitial anti-microbial genes. Previous study has indicated that the down-regulation of CXCL12 in CRC cell lines and primary tumor tissues may play a regulatory role in the initiation of CRC [59]. Experimentally, Wendt et al. reported that CXCL12 mRNA/protein is silenced by its promoter DNA hyper-methylation in primary colorectal tumor and cell lines [60]. The down-regulated CXCL12 is linked to tumor cells to resist anoikis, survive detachment, and circulate, since CXCL12 acts as a safeguard against metastasis by inducing anoikis [61]. These reports are consistent with our findings that *CXCL12* is hyper-methylated and down-regulated in COAD. Conversely, however, several studies have indicated that high levels of CXCL12 promote tumor growth, invasion, and poor prognosis in CRC [62]. CXCL12 and its receptor CXCR4 are crucial in the metastatic process of CRC [62]. The CXCL12/CXCR4 axis facilitates the migration and invasion of cancer cells to distant organs, particularly the liver and lungs, which are common sites of metastasis in CRC patients [63]. The CXCL12/CXCR4 interaction also promotes angiogenesis, the formation of new blood vessels, which is essential for tumor growth and providing nutrients to cancer cells [64]. CXCL12 contributes to the formation of a tumor-supportive micro-environment by recruiting immune cells, fibroblasts, and endothelial cells, which can aid in tumor growth and metastasis [62]. Due to its role in CRC progression and metastasis, CXCL12 and its receptor CXCR4 are considered potential therapeutic targets. Inhibiting the CXCL12/CXCR4 axis may provide a strategy to limit tumor growth and prevent metastasis in patients with CRC. Therefore, contradictory expression of CXCL12 might be related to early tumorigenesis and late metastasis in CRC.

hsa-miR-135b is an miRNA known for its roles in various cancers, acting either as an oncogene or a tumor suppressor depending on the context and tissue type. In colon cancer, hsa-miR-135b has been implicated in regulating several key genes involved in oncogenic pathways. hsa-miR-135b is known from several in vitro studies to directly down-regulate *CXCL12* [65]. In our research, it is also predicted that hsa-miR-135b negatively regulates *CXCL12* expression by binding to its mRNA and inhibiting its translation. This interaction can impact the CXCL12/CXCR4 signaling pathway, which is crucial for cancer cell migration and invasion. The dysregulation of hsa-miR-135b, leading to altered *CXCL12* expression, can significantly affect tumor behavior. Over-expression of hsa-miR-135b and subsequent down-regulation of *CXCL12* might reduce the chemotactic and angiogenic capabilities of cancer cells, potentially inhibiting metastasis.

CLDN1, *INHBA*, and *CXCL12* may serve as clinically relevant biomarkers in CRC, particularly as prognostic indicators linked to tumor progression, immune modulation, and metastatic potential. Over-expression of *CLDN1* has been associated with epithelial–mesenchymal transition, increased invasiveness, and metastatic behavior [66,67], as well as modulation of epithelial barrier permeability, suggesting a role in drug delivery and treatment efficacy—especially in ROS-based cancer therapies. Up-regulation of *INHBA* is linked to tumor proliferation, immune cell infiltration, and the development of an immuno-suppressive micro-environment [68,69]. Its protein product, Activin A, promotes fibrosis, angiogenesis, and immune evasion, contributing to chemoresistance and poor prognosis in colorectal and gastric cancers [70]. CXCL12, through interaction with CXCR4, regulates tumor cell survival, invasion, and immune suppression [64,71]. High *CXCL12* expression is associated with lymph node metastasis and poor survival, and the CXCL12/CXCR4 axis facilitates the recruitment of regulatory T-cells and M2 macrophages that support tumor immune escape [64,71].

Despite the comprehensive nature of our multi-omics analysis, several limitations should be noted. Our findings are entirely based on publicly available datasets and computational algorithms, and do not provide direct biological or clinical validation. Without experimental evidence, the mechanistic and causal roles of these key hub genes in CRC pathogenesis cannot be fully established. Therefore, rigorous experimental validation is urgently required. Future studies must include in-vitro and in-vivo assays—such as knock-down, over-expression, epigenetic editing of key hub genes, and tumor modeling—to confirm their biological function and oncogenic potential. These experiments are essential not only to substantiate our findings, but also to evaluate their suitability as therapeutic targets. Moreover, prospective clinical investigations are necessary to assess their prognostic and predictive power across treatment modalities, including chemotherapy, immunotherapy, and epigenetic therapy. Further studies should also examine the regulatory interactions of miR-375 and miR-135b through functional miRNA assays. Ultimately, integrating these biomarkers into non-invasive diagnostic platforms—such as exosomal miRNA profiling or ctDNA methylation panels—could enable real-time disease monitoring and individualized treatment strategies. Such follow-up studies are not optional but represent a critical next step for the clinical translation of bioinformatics-driven discoveries in CRC.

Clinical Recommendations

- *CLDN1* and *INHBA* are consistently over-expressed in COAD and are associated with poor prognosis and tumor progression, suggesting their potential role as negative prognostic biomarkers.
- *CXCL12* is down-regulated and epigenetically silenced in COAD and may be in-volved in early tumor suppression, offering insight into immune micro-environment modulation.
- Epigenetic regulation (via miRNAs and DNA methylation) plays a critical role in gene dysregulation in COAD and may represent therapeutic targets or predictive markers.
- Integrated multi-omics analysis improves the identification of functionally relevant and clinically applicable biomarkers for personalized treatment strategies in COAD.

5. Conclusions

Our integrative multi-omics analysis identified *CLDN1*, *INHBA*, and *CXCL12* as key biomarkers in COAD, regulated by both miRNAs and DNA methylation. These genes showed significant associations with tumor progression, immune infiltration, and patient prognosis. Our findings provide insight into the molecular landscape of COAD and suggest that multi-layered biomarkers may guide the development of personalized treatment strategies. Further experimental and clinical validation is warranted to translate these insights into therapeutic applications.

Author Contributions: Conceptualization, J.C., S.H.K. and T.H.K.; formal analysis, J.C. and T.H.K.; funding acquisition, T.H.K.; investigation, J.C.; methodology, J.C.; project administration, T.H.K.; supervision, T.H.K.; validation, S.H.K. and J.P.; visualization, J.C.; writing—original draft, J.C. and S.H.K.; writing—review and editing, J.C. and T.H.K. All authors have read and agreed to the published version of the manuscript.

Funding: This work was supported by the Basic Science Research Program of the National Research Foundation of Korea and funded by the Ministry of Science and Technology (RS-2024-00441029 and RS-2025-00513676), ICT Creative Consilience Program through the Institute of Information & Communications Technology Planning & Evaluation (IITP) grant funded by the Korea government (MSIT) (IITP-2025-RS-2020-II201819, 25%), Korea Health Technology R&D Project through the Korea Health Industry Development Institute (KHIDI), funded by the Ministry of Health & Welfare, Republic of Korea (RS-2022-KH129266). Additionally, this research was supported by a grant from the Korea University College of Medicine and Anam Hospital in Seoul, Republic of Korea.

Institutional Review Board Statement: Not applicable.

Informed Consent Statement: Not applicable.

Data Availability Statement: The data used in this study are available from the NCBI-GEO (https://www.ncbi.nlm.nih.gov/geo/, accessed on 15 January 2024), UALCAN (https://david.ncifcrf.gov, accessed on 5 February 2024), ENCORI/starBase (https://rnasysu.com/encori/, accessed on 19 February 2024), Single-Cell Portal database (https://singlecell.broadinstitute.org/single_cell, accessed on 12 February 2024), TIMER 2.0 (http://timer.cistrome.org/, accessed on 12 February 2024), Tumor, Normal, and Metastatic tissues plot.com database (https://tnmplot.com/, accessed on 19 February 2024), and KM plot database (https://kmplot.com/, accessed on 19 February 2024) databases, as well as, upon reasonable request, from the corresponding author.

Conflicts of Interest: The authors confirm that we have no conflicts of interest.

References

1. Xi, Y.; Xu, P. Global colorectal cancer burden in 2020 and projections to 2040. *Transl. Oncol.* **2021**, *14*, 101174. [CrossRef] [PubMed]
2. Marcellinaro, R.; Spoletini, D.; Grieco, M.; Avella, P.; Cappuccio, M.; Troiano, R.; Lisi, G.; Garbarino, G.M.; Carlini, M. Colorectal cancer: Current updates and future perspectives. *J. Clin. Med.* **2024**, *13*, 40. [CrossRef] [PubMed]
3. Yi, K.; Wu, J.; Tang, X.; Zhang, Q.; Wang, B.; Wang, F. Identification of a novel glycolysis-related gene signature for predicting the survival of patients with colon adenocarcinoma. *Scand. J. Gastroenterol.* **2022**, *57*, 214–221. [CrossRef] [PubMed]
4. Wang, Y.; Wang, J.; Gao, J.; Ding, M.; Li, H. The expression of SERPINE1 in colon cancer and its regulatory network and prognostic value. *BMC Gastroenterol.* **2023**, *23*, 33. [CrossRef]
5. Huang, H.; Xu, S.; Chen, A.; Li, F.; Wu, J.; Tu, X.; Hu, K. Identification of a 5-gene-based scoring system by WGCNA and LASSO to predict prognosis for rectal cancer patients. *Anal. Cell Pathol.* **2021**, *2021*, 6697407. [CrossRef]
6. Dayde, D.; Tanaka, I.; Jain, R.; Tai, M.C.; Taguchi, A. Predictive and prognostic molecular biomarkers for response to neoadjuvant chemoradiation in rectal cancer. *Int. J. Mol. Sci.* **2017**, *18*, 573. [CrossRef]
7. De Felice, F.; Crocetti, D.; Maiuri, V.; Parisi, M.; Marampon, F.; Izzo, L.; De Toma, G.; Musio, D.; Tombolini, V. Locally advanced rectal cancer: Treatment approach in elderly patients. *Curr. Treat. Options Oncol.* **2020**, *21*, 1. [CrossRef]
8. Ciardiello, F.; Ciardiello, D.; Martini, G.; Napolitano, S.; Tabernero, J.; Cervantes, A. Clinical management of metastatic colorectal cancer in the era of precision medicine. *CA Cancer J. Clin.* **2022**, *72*, 372–401. [CrossRef]
9. Dariya, B.; Aliya, S.; Merchant, N.; Alam, A.; Nagaraju, G.P. Colorectal cancer biology, diagnosis, and therapeutic approaches. *Crit. Rev. Oncog.* **2020**, *25*, 71–94. [CrossRef]
10. Ding, S.; Sun, X.; Zhu, L.; Li, Y.; Chen, W.; Shen, K. Identification of a novel immune-related prognostic signature associated with tumor microenvironment for breast cancer. *Int. Immunopharmacol.* **2021**, *100*, 108122. [CrossRef]
11. Luo, R.; Guo, W.; Wang, H. A comprehensive analysis of tumor microenvironment-related genes in colon cancer. *Clin. Transl. Oncol.* **2021**, *23*, 1769–1781. [CrossRef] [PubMed]
12. Babar, Q.; Saeed, A.; Tabish, T.A.; Sarwar, M.; Thorat, N.D. Targeting the tumor microenvironment: Potential strategy for cancer therapeutics. *Biochim. Biophys. Acta Mol. Basis Dis.* **2023**, *1869*, 166746. [CrossRef] [PubMed]

13. Peng, S.; Xiao, F.; Chen, M.; Gao, H. Tumor-microenvironment-responsive nanomedicine for enhanced cancer immunotherapy. *Adv. Sci.* **2022**, *9*, 2103836. [CrossRef] [PubMed]
14. Arneth, B. MicroRNAs in the tumor microenvironment. *Medicina* **2019**, *56*, 15. [CrossRef]
15. Yang, J.; Xu, J.; Wang, W.; Zhang, B.; Yu, X.; Shi, S. Epigenetic regulation in the tumor microenvironment: Molecular mechanisms and therapeutic targets. *Signal Transduct. Target. Ther.* **2023**, *8*, 210. [CrossRef]
16. Chen, S.; Shen, X. Long non-coding RNAs: Functions and mechanisms in colon cancer. *Mol. Cancer* **2020**, *19*, 167. [CrossRef]
17. Sean, D.; Meltzer, P.S. GEOquery: A bridge between the Gene Expression Omnibus (GEO) and BioConductor. *Bioinformatics* **2007**, *23*, 1846–1847.
18. Sherman, B.T.; Huang, D.W.; Tan, Q.; Guo, Y.; Bour, S.; Liu, D.; Stephens, R.; Baseler, M.W.; Lane, C.H.; Lempicki, R.A. DAVID Knowledgebase: A gene-centered database integrating heterogeneous gene annotation resources to facilitate high-throughput gene functional analysis. *BMC Bioinform.* **2007**, *8*, 426. [CrossRef]
19. Szklarczyk, D.; Gable, A.L.; Nastou, K.C.; Lyon, D.; Kirsch, R.; Pyysalo, S.; Doncheva, N.T.; Legeay, M.; Fang, T.; Bork, P.; et al. The STRING database in 2021: Customizable protein-protein networks, and functional characterization of user-uploaded gene/measurement sets. *Nucleic Acids Res.* **2021**, *49*, D605–D612. [CrossRef]
20. Chandrashekar, D.S.; Bashel, B.; Balasubramanya, S.A.H.; Creighton, C.J.; Ponce-Rodriguez, I.; Chakravarthi, B.V.S.K.; Varambally, S. UALCAN: A portal for facilitating tumor subgroup gene expression and survival analyses. *Neoplasia* **2017**, *19*, 649–658. [CrossRef]
21. Li, J.H.; Liu, S.; Zhou, H.; Qu, L.H.; Yang, J.H. StarBase v2.0: Decoding miRNA-ceRNA, miRNA-ncRNA, and protein-RNA interaction networks from large-scale CLIP-Seq data. *Nucleic Acids Res.* **2014**, *42*, 92–97. [CrossRef] [PubMed]
22. Shi, C.Y.; Kingston, E.R.; Kleaveland, B.; Lin, D.H.; Stubna, M.W.; Bartel, D.P. The ZSWIM8 ubiquitin ligase mediates target-directed microRNA degradation. *Science* **2020**, *370*, 6523. [CrossRef]
23. Egger, G.; Aparicio, A.; Jones, P.A.; Liang, G. Epigenetics in human diseases and prospects of epigenetic therapy. *Nature* **2004**, *429*, 457–463. [CrossRef]
24. Tarhan, L.; Bistline, J.; Chang, J.; Galloway, B.; Hanna, E.; Weitz, E. Single cell portal: An interactive home for single-cell genomics data. *bioRxiv* **2023**. bioRxiv: 2023.07.13.548886.
25. Li, T.; Fu, J.; Zeng, Z.; Cohen, D.; Li, J.; Chen, Q.; Li, B.; Liu, X.S. TIMER2.0 for analysis of tumor-infiltrating immune cells. *Nucleic Acids Res.* **2020**, *48*, W509–W514. [CrossRef]
26. Bartha, Á.; Győrffy, B. Tnmplot.Com: A web tool for the comparison of gene expression in normal, tumor and metastatic tissues. *Int. J. Mol. Sci.* **2021**, *22*, 2622. [CrossRef]
27. Győrffy, B. Transcriptome-level discovery of survival-associated biomarkers and therapy targets in non-small-cell lung cancer. *Br. J. Pharmacol.* **2024**, *181*, 362–374. [CrossRef] [PubMed]
28. Li, J.; Miao, B.; Wang, S.; Dong, W.; Xu, H.; Si, C.; Wang, W.; Duan, S.; Lou, J.; Bao, Z.; et al. Hiplot: A comprehensive and easy-to-use web service for boosting publication-ready biomedical data visualization. *Brief. Bioinform.* **2022**, *23*, bbac261. [CrossRef]
29. Wang, X.; Duanmu, J.; Fu, X.; Li, T.; Jiang, Q. Analyzing and validating the prognostic value and mechanism of colon cancer immune microenvironment. *J. Transl. Med.* **2020**, *18*, 324. [CrossRef]
30. Neophytou, C.M.; Panagi, M.; Stylianopoulos, T.; Papageorgis, P. The role of tumor microenvironment in cancer metastasis: Molecular mechanisms and therapeutic opportunities. *Cancers* **2021**, *13*, 2053. [CrossRef]
31. Pajares, M.J.; Alemany-cosme, E.; Goñi, S.; Bandres, E.; Palanca-Ballester, C.; Sandoval, J. Epigenetic regulation of microRNAs in cancer: Shortening the distance from bench to bedside. *Int. J. Mol. Sci.* **2021**, *22*, 7350. [CrossRef] [PubMed]
32. Aure, M.R.; Fleischer, T.; Bjørklund, S.; Ankill, J.; Castro-Mondragon, J.A.; Bathen, T.F.; Borgen, E.; Engebråten, O.; Hartman-Johnsen, O.J.; Garred, Ø.; et al. Crosstalk between microRNA expression and DNA methylation drives the hormone-dependent phenotype of breast cancer. *Genome Med.* **2021**, *13*, 72. [CrossRef]
33. Okayama, H.; Schetter, A.J.; Harris, C.C. MicroRNAs and inflammation in the pathogenesis and progression of colon cancer. *Dig. Dis.* **2012**, *30*, 9–15. [CrossRef] [PubMed]
34. Zhang, L.; Fan, X.M. The pathological role of microRNAs and inflammation in colon carcinogenesis. *Clin. Res. Hepatol. Gastroenterol.* **2015**, *39*, 174–179. [CrossRef] [PubMed]
35. Okano, M.; Yamamoto, H.; Ohkuma, H.; Kano, Y.; Kim, H.; Nishikawa, S.; Konno, M.; Kawamoto, K.; Haraguchi, N.; Takemasa, I.; et al. Significance of INHBA expression in human colorectal cancer. *Oncol. Rep.* **2013**, *30*, 2903–2908. [CrossRef]
36. Seder, C.W.; Hnrtojo, W.; Lin, L.; Silvers, A.L.; Wang, Z.; Thomas, D.G.; Giordano, T.J.; Chen, G.; Chang, A.C.; Orringer, M.B.; et al. Upregulated INHBA expression may promote cell proliferation and is associated with poor survival in lung adenocarcinoma. *Neoplasia* **2009**, *11*, 388–396. [CrossRef]
37. Katayama, Y.; Oshima, T.; Sakamaki, K.; Aoyama, T.; Sato, T.; Masudo, K.; Shiozawa, M.; Yoshikawa, T.; Rino, Y.; Imada, T.; et al. Clinical significance of INHBA gene expression in patients with gastric cancer who receive curative resection followed by adjuvant s-1 chemotherapy. *In Vivo* **2017**, *31*, 565–571.

38. Seder, C.W.; Hartojo, W.; Lin, L.; Silvers, A.L.; Wang, Z.; Thomas, D.G.; Giordano, T.J.; Chen, G.; Chang, A.C.; Orringer, M.B.; et al. INHBA overexpression promotes cell proliferation and may be epigenetically regulated in esophageal adenocarcinoma. *J. Thorac. Oncol.* **2009**, *4*, 455–462. [CrossRef]
39. Wildi, S.; Kleeff, J.; Maruyama, H.; Maurer, C.A.; Büchler, M.W.; Korc, M. Overexpression of activin A in stage IV colorectal cancer. *Gut* **2001**, *49*, 409–417. [CrossRef]
40. Dean, M.; Davis, D.A.; Burdette, J.E. Activin A stimulates migration of the fallopian tube epithelium, an origin of high-grade serous ovarian cancer, through non-canonical signaling. *Cancer Lett.* **2017**, *391*, 114–124. [CrossRef]
41. Lee, H.Y.; Li, C.C.; Huang, C.N.; Li, W.M.; Yeh, H.C.; Ke, H.L.; Yang, K.F.; Liang, P.I.; Li, C.F.; Wu, W.J. INHBA overexpression indicates poor prognosis in urothelial carcinoma of the urinary bladder and upper tract. *J. Surg. Oncol.* **2015**, *111*, 414–422. [CrossRef] [PubMed]
42. Guo, J.; Liu, Y. INHBA promotes the proliferation, migration, and invasion of colon cancer cells through the upregulation of VCAN. *J. Int. Med. Res.* **2021**, *49*, 03000605211014998. [CrossRef] [PubMed]
43. Wang, Y.; Tang, Q.; Li, M.; Jiang, S.; Wang, X. MicroRNA-375 inhibits colorectal cancer growth by targeting PIK3CA. *Biochem. Biophys. Res. Commun.* **2014**, *444*, 199–204. [CrossRef]
44. Xu, L.; Wen, T.; Liu, Z.; Xu, F.; Yang, L.; Liu, J.; Feng, G.; An, G. MicroRNA-375 suppresses human colorectal cancer metastasis by targeting Frizzled 8. *Oncotarget* **2016**, *7*, 40644–40656. [CrossRef]
45. Han, K.H.; Cho, H.; Han, K.R.; Mun, S.K.; Kim, Y.K.; Park, I.; Chang, M. Role of microRNA-375-3p-mediated regulation in tinnitus development. *Int. J. Mol. Med.* **2021**, *48*, 136. [CrossRef]
46. Kang, W.; Huang, T.; Zhou, Y.; Zhang, J.; Lung, R.W.M.; Tong, J.H.M.; Chan, A.W.H.; Zhang, B.; Wong, C.C.; Wu, F.; et al. MiR-375 is involved in the Hippo pathway by targeting the YAP1/TEAD4-CTGF axis in the gastric carcinogenesis article. *Cell Death Dis.* **2018**, *9*, 92. [CrossRef]
47. Łukaszewicz-Zając, M.; Mroczko, B. Claudins-promising biomarkers for selected gastrointestinal (GI) malignancies? *Cancers* **2023**, *16*, 152. [CrossRef]
48. Morin, P.J. Claudin proteins in human cancer: Promising new targets for diagnosis and therapy. *Cancer Res.* **2005**, *65*, 9603–9606. [CrossRef] [PubMed]
49. Ouban, A. Claudin-1 role in colon cancer: An update and a review. *Histol. Histopathol.* **2018**, *33*, 1013–1019.
50. Huang, J.; Li, J.; Qu, Y.; Zhang, J.; Zhang, L.; Chen, X.; Liu, B.; Zhu, Z. The expression of Claudin 1 correlates with β-catenin and is a prognostic factor of poor outcome in gastric cancer. *Int. J. Oncol.* **2014**, *44*, 1293–1301. [CrossRef]
51. De Vicente, J.C.; Fernández-Valle, Á.; Vivanco-Allende, B.; Rodríguez Santamarta, T.; Lequerica-Fernández, P.; Hernández-Vallejo, G.; Allonca-Campa, E. The prognostic role of claudins -1 and -4 in oral squamous cell carcinoma. *Anticancer Res.* **2015**, *35*, 2949–2960.
52. De Oliveira, S.S.; De Oliveira, I.M.; De Souza, W.; Morgado-Díaz, J.A. Claudins upregulation in human colorectal cancer. *FEBS Lett.* **2005**, *579*, 6179–6185. [CrossRef] [PubMed]
53. Kinugasa, T.; Akagi, Y.; Shirouzu, K. Claudin-1 protein is a major factor involved in the tumorigenesis of colorectal. *Gastroenterology* **2009**, *136*, A-320. [CrossRef]
54. Dhawan, P.; Singh, A.B.; Deane, N.G.; No, Y.R.; Shiou, S.R.; Schmidt, C.; Neff, J.; Washington, M.K.; Beauchamp, R.D. Claudin-1 regulates cellular transformation and metastatic behavior in colon cancer. *J. Clin. Investig.* **2005**, *115*, 1765–1776. [CrossRef]
55. Kinugasa, T.; Akagi, Y.; Ochi, T.; Tanaka, N. Increased claudin-1 protein expression in hepatic metastatic lesions of colorectal cancer. *Anticancer Res.* **2012**, *32*, 2309–2314. [PubMed]
56. Abdelzaher, E.; Rizk, A.M.; Bessa, S.S.; Omer, K.M. Predictive value of immunohistochemical expression of claudin-1 in colonic carcinoma. *J. Egypt. Natl. Cancer Inst.* **2011**, *23*, 123–131. [CrossRef]
57. Shibutani, M.; Noda, E.; Maeda, K.; Nagahara, H.; Ohtani, H.; Hirakawa, K. Low expression of Claudin-1 and presence of poorly differentiated tumor clusters correlate with poor prognosis in colorectal cancer. *Anticancer Res.* **2013**, *33*, 3301–3306.
58. Nishikawa, E.; Osada, H.; Okazaki, Y.; Arima, C.; Tomida, S.; Tatematsu, Y.; Taguchi, A.; Shimada, Y.; Yanagisawa, K.; Yatabe, Y.; et al. MiR-375 is activated by ASH1 and inhibits YAP1 in a lineage-dependent manner in lung cancer. *Cancer Res.* **2011**, *71*, 6165–6173. [CrossRef]
59. Hozhabri, H.; Lashkari, A.; Razavi, S.M.; Mohammadian, A. Integration of gene expression data identifies key genes and pathways in colorectal cancer. *Med. Oncol.* **2021**, *38*, 7. [CrossRef]
60. Wendt, M.K.; Johanesen, P.A.; Kang-Decker, N.; Binion, D.G.; Shah, V.; Dwinell, M.B. Silencing of epithelial CXCL12 expression by DNA hypermethylation promotes colonic carcinoma metastasis. *Oncogene* **2006**, *25*, 4986–4997. [CrossRef]
61. Drury, L.J.; Wendt, M.K.; Dwinell, M.B. CXCL12 chemokine expression and secretion regulates colorectal carcinoma cell anoikis through Bim-mediated intrinsic apoptosis. *PLoS ONE* **2010**, *5*, e12895. [CrossRef] [PubMed]
62. Bocchi, M.; de Sousa Pereira, N.; de Oliveira, K.B.; Amarante, M.K. Involvement of CXCL12/CXCR4 axis in colorectal cancer: A mini-review. *Mol. Biol. Rep.* **2023**, *50*, 6233–6239. [CrossRef]
63. Janssens, R.; Struyf, S.; Proost, P. The unique structural and functional features of CXCL12. *Cell. Mol. Immunol.* **2018**, *15*, 299–311. [CrossRef]

64. Khare, T.; Bissonnette, M.; Khare, S. Cxcl12-cxcr4/cxcr7 axis in colorectal cancer: Therapeutic target in preclinical and clinical studies. *Int. J. Mol. Sci.* **2021**, *22*, 7371. [CrossRef] [PubMed]
65. Tamaru, S.; Mizuno, Y.; Tochigi, H.; Kajihara, T.; Okazaki, Y.; Okagaki, R.; Kamei, Y.; Ishihara, O.; Itakura, A. MicroRNA-135b suppresses extravillous trophoblast-derived HTR-8/SVneo cell invasion by directly downregulating CXCL12 under low oxygen conditions. *Biochem. Biophys. Res. Commun.* **2015**, *461*, 421–426. [CrossRef] [PubMed]
66. Primeaux, M.; Liu, X.; Gowrikumar, S.; Fatima, I.; Fisher, K.W.; Bastola, D.; Vecchio, A.J.; Singh, A.B.; Dhawan, P. Claudin-1 interacts with EPHA2 to promote cancer stemness and chemoresistance in colorectal cancer. *Cancer Lett.* **2023**, *579*, 216479. [CrossRef]
67. Cherradi, S.; Garambois, V.; Marines, J.; Andrade, A.F.; Fauvre, A.; Morand, O.; Fargal, M.; Mancouri, F.; Ayrolles-Torro, A.; Vezzo-Vié, N.; et al. Improving the response to oxaliplatin by targeting chemotherapy-induced CLDN1 in resistant metastatic colorectal cancer cells. *Cell Biosci.* **2023**, *13*, 72. [CrossRef]
68. Zhang, X.; Wu, T.; Qin, R.; Cai, X.; Zhou, Y.; Wang, X.; Shang, Z.; Li, G.; Yang, R.; Dong, C.; et al. The New Role of HNF1A-NAS1/miR-214/INHBA Signaling Axis in Colorectal Cancer. *Front. Biosci. Landmark* **2023**, *28*, 301. [CrossRef]
69. Lin, H.; Hong, Y.G.; Zhou, J.D.; Gao, X.H.; Yuan, P.H.; Xin, C.; Huang, Z.P.; Zhang, W.; Hao, L.Q.; Hou, K.Z. LncRNA INHBA-AS1 promotes colorectal cancer cell proliferation by sponging miR-422a to increase AKT1 axis. *Eur. Rev. Med. Pharmacol. Sci.* **2020**, *24*, 9940–9948.
70. Qiu, S.; Li, B.; Xia, Y.; Xuan, Z.; Li, Z.; Xie, L.; Gu, C.; Lv, J.; Lu, C.; Jiang, T.; et al. CircTHBS1 drives gastric cancer progression by increasing INHBA mRNA expression and stability in a ceRNA- and RBP-dependent manner. *Cell Death Dis.* **2022**, *13*, 266. [CrossRef]
71. Wang, D.; Wang, X.; Song, Y.; Si, M.; Sun, Y.; Liu, X.; Cui, S.; Qu, X.; Yu, X. Exosomal miR-146a-5p and miR-155-5p promote CXCL12/CXCR7-induced metastasis of colorectal cancer by crosstalk with cancer-associated fibroblasts. *Cell Death Dis.* **2022**, *13*, 380. [CrossRef] [PubMed]

biomedicines

MDPI

Review

G-Quadruplexes in Tumor Immune Regulation: Molecular Mechanisms and Therapeutic Prospects in Gastrointestinal Cancers

Yunxia Zhou †, Difei Xu †, Ying Zhang * and Huaixiang Zhou *

Tomas Lindahl Nobel Laureate Laboratory, The Seventh Affiliated Hospital, Sun Yat-Sen University, Shenzhen 518107, China

* Correspondence: zhangy856@mail.sysu.edu.cn (Y.Z.); zhouhuaixiang@sysush.com (H.Z.)

† These authors contributed equally to this work.

Abstract: G-quadruplex (G4) is a noncanonical nucleic acid secondary structure self-assembled by guanine-rich sequences. Recent studies have not only revealed the key role of G4 in gene regulation, DNA replication, and telomere maintenance but also showed that it plays a core role in regulating the tumor immune microenvironment. G4 participates in tumor immune escape and the inhibition of immune response by regulating immune checkpoint molecules, cytokine expression, immune cell function, and their interaction network, thus significantly affecting the effect of tumor immunotherapy. This article systematically reviews the molecular mechanism of G4 in tumor immune regulation, especially gastrointestinal tumors, and explores the potential and application prospects of G4-targeted drug strategies in improving anti-tumor immunotherapy.

Keywords: G-quadruplexes; tumor immune microenvironment; gastrointestinal tumors; treatment strategies

1. Introduction

The discovery and study of G4 structures began in the 1960s, and with a deeper understanding of DNA and RNA molecular structures, G4 has emerged as an important atypical nucleic acid structure that plays a critical role in gene regulation, cell cycle, DNA repair, and telomere maintenance [1–8]. In cancer, neurodegenerative diseases, and viral infections, G4 structures have increasingly become the focus of emerging targeted therapeutic strategies [9–12]. G4 structures are abundant in proto-oncogene regions of tumor cells, where they are closely related to the transcriptional regulation of oncogenes, tumor immune escape, tumor microenvironment alterations, and therapeutic resistance [9,13–15]. Studies have shown that G4 can participate in the immune escape process of tumors by regulating immune checkpoint genes (e.g., PD-L1), inflammatory cytokines, and immune cell functions, thus influencing the effectiveness of immunotherapy [16–18]. Furthermore, G4 structures are closely related to DNA replication, genome stability, and telomere maintenance, making them potential targets for anticancer therapy [7,19,20].

As research into G4 structure and function deepens, a growing number of G4-targeted drugs—including G4 stabilizers, unwinders, binders, and G4-based immunotherapy strategies—have been developed [18,21–24]. These drugs not only show promise in cancer treatment but may also offer novel ideas for treating other diseases. Although the function and mechanism of G4 have been extensively studied in recent years, challenges remain regarding their specific roles in different diseases, the efficacy of targeting strategies, and the clinical translation of these drugs [25–28]. Particularly in cancer immunotherapy, enhancing the effectiveness of immune checkpoint inhibitors through G4 targeting is a major area of current research [29–33]. This review systematically summarizes the mechanism of action of G4 in tumor immune regulation and gastrointestinal tumors, and explores the therapeutic potential and future application direction of drugs targeting G4, in order to provide theoretical support for cancer treatment strategies.

2. Key Features of G-Quadruplexes

2.1. Structure of G-Quadruplexes

G4 is a specific nucleic acid secondary structure formed by guanine-rich sequences through Hoogsteen hydrogen bonds [34]. Its basic structural unit, the G-tetrad, is formed by four guanine molecules linked in a planar arrangement, which then stack via π–π interactions to form the quadruplex [35,36]. The stability of G4 structures is influenced by monovalent cations such as K^+ and Na^+, with K^+ providing stronger stabilization due to its optimal ionic radius and coordination properties, as well as its high physiological concentration [37,38]. G4 can adopt various conformations, including parallel, antiparallel, and hybrid forms [36,39–42] (Table 1).

Table 1. The key characteristics of quadruplex.

Features	Description
Basic structure	G-tetrad is formed by Hoogsteen hydrogen bonding
Stability Factors	Stabilized by monovalent cations (K^+, Na^+); magnesium ions can further enhance stability
Structural Types	Parallel, antiparallel, and hybrid; can form single-stranded, double-stranded, or multi-stranded structures
Genomic Localization	Located in telomeres, promoters, enhancers, replication origins, and non-coding RNA regions

2.2. Distribution of G-Quadruplexes

G4 structures are widely distributed in functional genomic regions:

Telomeres: Telomeric DNA, rich in guanine repeats, readily forms G4 structures that affect telomere maintenance and telomerase activity [7,43].

Promoters: G4 structures are enriched in the promoters of many oncogenes (e.g., MYC, KRAS, BCL-2) and play a role in transcriptional regulation [26,37,44–46].

Enhancers and Regulatory Elements: In gene regulatory regions such as super-enhancers, G4 structures can recruit specific proteins to modulate gene expression [47–49].

Replication Origins: G4 influences the initiation of DNA replication and collaborates with chromatin remodeling factors [50,51].

Non-coding RNAs: G4 structures have been found in long non-coding RNAs (lncRNAs) and microRNA precursors, affecting RNA processing and function [52–57] (Table 1).

2.3. Conformational Diversity of G-Quadruplex Structures

Recent advances in high-throughput G4-mapping techniques, such as G4-seq, rG4-seq, and BG4 ChIP-seq, have revealed that G4 structures are not uniformly distributed across the genome and exhibit context-dependent conformational preferences [58–60]. While parallel G4 conformations are frequently observed in promoter regions and CpG islands, likely due to their short loop lengths and higher thermodynamic stability, antiparallel and hybrid G4s are more commonly found in telomeres and non-coding regions, reflecting the influence of local sequence context and chromatin structure [2,33]. Moreover, the folding topology of G4s can directly impact their biological functions. For instance, parallel G4s in oncogene promoters (e.g., MYC, KRAS) act as transcriptional repressors by interfering with transcription factor binding or RNA polymerase progression [61], while telomeric hybrid G4s serve as binding platforms for telomere-associated proteins such as POT1 and TRF2, playing critical roles in telomere protection and genome stability [62]. In the 5′-untranslated regions (5′-UTRs) of mRNAs, G4 structures can adopt looped or bulged forms that modulate cap-dependent translation or facilitate ribosome stalling, thereby influencing protein synthesis in a gene-specific manner [63]. Importantly, chromatin environment, supercoiling, and RNA/DNA interactions further modulate G4 folding dynamics and topology in vivo, resulting in functional diversification even among structurally similar motifs [64]. Therefore, G4 structures exhibit conformational diversity across different genomic regions, with their folding topology closely linked to biological functions and jointly regulated by sequence context and chromatin environment.

3. Functions of G-Quadruplexes

3.1. G-Quadruplexes in Gene Regulation

DNA Replication: G4 can impede DNA replication by causing replication fork stalling. DNA helicases such as BLM and WRN help resolve G4 structures to ensure smooth replication [65].

Transcriptional Regulation: G4 can regulate gene expression by affecting RNA polymerase II binding, obstructing transcription factor binding, or recruiting G4-binding proteins (G4BP) to either activate or repress transcription [66–68]. For example, the G4 in the MYC promoter acts as a transcriptional repressor, and small molecule stabilizers like Pyridostatin can downregulate MYC expression [69].

RNA Processing and Translation Regulation: G4 structures in mRNA, particularly in the 5′- and 3′-UTRs, can modulate splicing, translation, and mRNA stability. G4 in NRAS and VEGF mRNAs has been shown to inhibit translation, underscoring its regulatory role in cancer-related genes [70–73].

Telomere Maintenance: Telomere maintenance is essential for genome stability and cellular lifespan, especially in rapidly dividing tumor cells [74,75]. G4 structures in telomeric regions can inhibit telomerase activity, thereby controlling telomere length and serving as potential targets for anticancer therapy [7,43,76].

Genome Stability: Class switch recombination (CSR) is a biological mechanism by which B cells change the isotype of the antibody they produce (e.g., from IgM to IgG, IgA, or IgE) without altering antigen specificity. This process is crucial for tailoring immune responses and is tightly regulated by activation-induced cytidine deaminase (AID) and associated DNA repair mechanisms [77]. G4 is important for the B-cell lineage and is found in high abundance in Ig variable (V) genes [78] and in so-called "switch" (S) regions targeted by the CSR process [79]. At these locations, G4 plays a regulatory role but can also compromise genomic stability by initiating double-strand breaks (DSBs) and translocations (Figure 1).

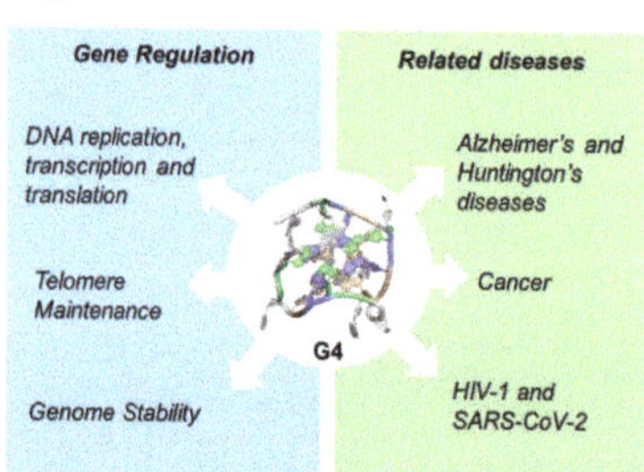

Figure 1. Functions of G-quadruplexes.

3.2. G-Quadruplexes in Diseases

Cancer: The presence of G4 in oncogene promoters and telomeres affects oncogene expression, telomere maintenance, and DNA damage responses. Thus, G4 stabilizers (e.g., CX-5461, Pyridostatin) are being explored as novel anticancer agents [16,80–82].

Neurodegenerative Diseases: Abnormal G4-mediated RNA regulation has been linked to neurodegenerative disorders such as Alzheimer's and Huntington's diseases. For example, G4 structures in the FMR1 gene affect mRNA translation and are associated with Fragile X syndrome [83–85].

Viral Infections: G4 plays a role in viral genomes and host antiviral responses. For example, G4 in the HIV-1 and SARS-CoV-2 genomes affects viral RNA translation and replication, suggesting that targeting G4 could be a promising antiviral strategy [86–89] (Figure 1).

4. G-Quadruplexes in Cancer Immunoregulation

Recent studies have found that G4 plays an important regulatory role in the cancer immune microenvironment, mainly mediating tumor cell immune escape and immunosuppression by affecting immune checkpoint molecule expression, cytokine regulation, immune cell function, and molecular interaction networks [4,16,18,30,90] (Table 2).

Table 2. Mechanisms of G-quadruplexes in cancer immune regulation.

Mechanism	Molecular Targets	Regulation	Biological Effects	Ref.
Immune Checkpoint Gene Regulation	PD-L1	G4 promotes transcription	Reduces immune suppressive molecule expression, enhances T cell cytotoxicity	[16]
mRNA Stability and Translation Regulation	IL-6, IL-8, IFN-γ, TNF-α	G4 enhances or inhibits mRNA stability	Affects cytokine secretion, regulates inflammation and immune microenvironment	[91]
Immune Cell Function Regulation	T cells, DCs, macrophages	G4 regulates DNA damage response and epigenetic modifications	Influences immune cell differentiation, activation, and antigen presentation capacity	[92–96]
Molecular Interaction Network Influence	lncRNA, DNA methylation	G4 mediates gene regulatory networks	Affects the epigenetic regulation of immune genes, shapes tumor microenvironment features	[97–99]

4.1. Transcriptional Regulation of Immune Checkpoint Genes

G-quadruplex (G4) structures have emerged as important epigenetic regulators capable of indirectly modulating PD-L1 transcription, with evidence suggesting context-dependent and sometimes opposing regulatory effects. One study found that certain guanine-rich oligonucleotides, such as nCpG-6-PTO, can form G4 structures and significantly downregulate PD-L1 expression in melanoma cells [100]. However, another study demonstrated that G4 stabilizers, such as CX-5461, can indirectly upregulate PD-L1 expression by activating the cGAS–STING–IFN signaling pathway [16]. Therefore, the regulatory effect of G4 stabilizers on PD-L1 is context-dependent and may involve immune signaling-mediated indirect upregulation. These findings highlight the importance of carefully selecting G4 ligands and optimizing dosing strategies when designing combination therapies with immune checkpoint inhibitors.

4.2. Regulation of mRNA Translation and Cytokine Secretion

Beyond their effects on transcription, G4 structures also regulate the translation of cytokines and immune-related molecules through the formation of G4 structures in the 5′-untranslated regions (5′-UTR) of mRNAs [91]. Cytokines such as IL-6, IL-8, and VEGF are critical for immune cell recruitment and activation within the tumor microenvironment [17,101,102]. G4 structures in the 5′-UTRs of these cytokine mRNAs have been shown to influence ribosome binding, thereby modulating translation efficiency [103]. For example, IL-6 and TNF-α exhibit increased translation when G4 structures are disrupted by G4 unwinders, suggesting that G4 stabilization could prevent the upregulation of pro-inflammatory cytokines, thereby influencing immune cell recruitment and activation [104]. The regulation of these cytokines by G4 structures in immune cells has profound implications for immune modulation. Disruption of G4 structures with specific G4 unwinders could therefore alter the inflammatory response and potentially enhance anti-tumor immunity. By modulating cytokine secretion, G4 structures might not only affect the immune landscape but also influence tumor progression, immune resistance, and therapeutic responses.

4.3. Immune Cell Function and Tumor Microenvironment

In addition to modulating gene expression at the transcriptional and translational levels, G4 structures also directly influence immune cell function, particularly in T cells, dendritic cells (DCs), and macrophages. These immune cells play pivotal roles in the anti-tumor immune response, and G4 may influence their differentiation, activation, and apoptosis [92–94]. For example, in T cells and DCs, G4 may affect their differentiation and activation by regulating DNA damage response (DDR), chromatin remodeling, and epigenetic modifications [95]. Furthermore, G4 structures are involved in the regulation of macrophage polarization. Tumor-associated macrophages (TAMs) often exhibit an immunosuppressive phenotype that supports tumor progression. Targeting G4 structures in these immune cells could potentially shift the macrophage phenotype toward an anti-tumor state, enhancing the overall immune response within the tumor [96]. Further exploration of this mechanism may provide a new theoretical basis for the application of G4-targeted drugs in enhancing anti-tumor immune response.

4.4. Molecular Interactions and Epigenetic Regulation

G4 may also work in synergy with epigenetic factors such as non-coding RNA (such as long non-coding RNA, lncRNA) and DNA methylation to further affect gene expression [97,98]. For example, in the regulatory axis (G4–lncRNA–immune gene) in the tumor microenvironment, G4 may act as a signal regulation switch to determine the interaction between tumor cells and immune cells [99]. Therefore, studying how G4 cooperates with these molecular networks to regulate tumor immune responses will help develop more precise targeted intervention strategies.

5. Therapeutic Potential of G-Quadruplex Regulation in Gastrointestinal Tumors

As an important secondary structure of gene regulation, G-quadruplex (G4) shows unique biological significance in gastrointestinal diseases, especially gastrointestinal tumors and infectious diseases. Studies in recent years have shown that G4 structure is widely distributed in the promoter region and 5′-UTR regions of oncogenes, and its formation is closely related to the regulation of gene expression [105]. In gastrointestinal stromal tumors (GIST), abnormal expression of the c-KIT gene is a key factor in tumor occurrence and development, and the G4 structure in the c-KIT promoter region has become a potential regulatory target [106,107]. By designing specific small molecules to stabilize the G4 structure, it is not only possible to inhibit the transcription of the c-KIT gene but also to overcome the treatment dilemma caused by kinase inhibitor resistance.

The formation of drug resistance is a major challenge in the treatment of gastrointestinal tumors, and the existence of G4 structure provides a new idea for revealing the mechanism of drug resistance. Studies have found that BCL-2, an anti-apoptotic gene, plays an important role in the drug resistance of GIST cells, and its G4 structure in the 5′-UTR regions may affect the expression level of BCL-2. Certain small molecule G4 binders can specifically target this structure, reduce BCL-2 expression, and thus restore cancer cells' sensitivity to targeted therapy. In addition, G4 is widely involved in the maintenance of DNA replication and genomic stability [106]. Excessive formation in gastrointestinal tumor cells may lead to blocked DNA replication, thereby affecting the proliferation ability of tumor cells. Therefore, the regulation of G4 structure not only plays an important role in the regulation of oncogene expression but also may become a new strategy to affect the growth of tumor cells.

The impact of G4 structure is not limited to host cells but also plays a key role in pathogens such as Helicobacter pylori. The latest study found that highly conserved G4 structures exist in the nickel transport-related genes (nixA, niuB1, niuB2, and niuD) of the pathogen [108]. These genes regulate the uptake of nickel ions in the host by Helicobacter pylori, and nickel is a key auxiliary factor in maintaining the function of Helicobacter pylori urease and hydrogenase [108]. Therefore, the presence of these G4 structures may affect the survival ability of Helicobacter pylori and provide new targets for anti-infection treatment. Small molecule G4 binders can not only be used to target the G4 structure of oncogenes, but also may stabilize the G4 structure in pathogens, interfere with the expression of their key genes, and ultimately weaken their pathogenicity. This discovery broadens the scope of the application of the G4 structure in disease treatment, making it not only an anti-tumor target but also a new idea for antibacterial therapy.

Targeting the G-quadruplex (G4) structure has become an emerging strategy for the treatment of gastrointestinal tumors. Studies have shown that the G4 structure plays a key role in oncogene expression, drug resistance regulation, and immune escape. Based on this, the development of small molecules or other therapeutic methods that can regulate the G4 structure is expected to provide new treatment options for patients with gastrointestinal tumors.

6. G-Quadruplex-Targeted Therapeutic Strategies

The development of therapeutic strategies targeting G4 has gained significant attention in recent years due to their critical role in regulating gene expression and their involvement in cancer progression, immune regulation, and other diseases. G4 structures have become promising targets for both anticancer therapy and immune modulation, with various approaches focusing on stabilizing or unwinding G4 structures to regulate gene expression and enhance immune responses. This section explores the different G4-targeted therapeutic strategies, including G4 stabilizers, G4 unwinders, combination therapies, and multifunctional nanomedicines (Table 3).

Table 3. G-quadruplex-targeted therapeutic strategies.

Strategy	Representative Drugs	Mechanism of Action	Potential Advantages	Challenges and Limitations	Ref
G4 Stabilizers	BRACO-19, TMPyP4, CX-5461, Pyridostatin	Stabilize G4 structures and inhibit target gene transcription (e.g., c-MYC, PD-L1)	Reduce tumor immune escape, enhance T cell cytotoxicity	Low selectivity, may affect normal cell gene expression	[109–111]
G4 Unwinders	Pif1 DNA helicase, 2′-F C3	Promoting the dissociation of G4 structures to prevent replication fork stalling and DNA breakage	Without Inducing DNA Damage	Potential for drug resistance or non-specific effects	[24,112]
Combination Immunotherapy	G4 Stabilizer + PD-1 Antibody	Modulate G4 to reduce PD-L1 while blocking the PD-1/PD-L1 pathway	Enhance the efficacy of immune checkpoint inhibitors	Requires optimization of dose matching and immune tolerance risks	[16,113,114]
Nanomedicine Delivery	G4-targeted Nanocarriers	Deliver G4 modulators to improve drug selectivity and stability	Reduce systemic toxicity, improve tumor targeting	Delivery system still needs optimization, significant clinical translation challenges	[115–117]

6.1. G4 Stabilizers

G4 stabilizers are small molecules that enhance the formation and stability of G4 structures, particularly in promoter regions of oncogenes, immune checkpoint genes, and other critical regulatory sites [16,58]. By stabilizing G4 structures, these agents can inhibit the transcription of target genes, suppress tumor growth, and enhance the immune response [18,118]. Several G4 stabilizers, such as CX-5461 (Pidnarulex), Pyridostatin, BRACO-19, and TMPyP4, have been identified and shown to selectively bind to G4 structures, blocking the activity of transcription factors or RNA polymerase [109–111]. The FDA has granted Fast Track designation to CX-5461 as a potential treatment option for breast and ovarian cancer patients harboring mutations in BRCA1/2, PALB2, or other homologous recombination deficiencies [119]. A G4-stabilizing compound, CX-5461 is designed to stabilize the folded conformation and synergize with homologous recombination (HR) repair pathway defects; this prevents DNA breaks at replication forks and leads to cancer death [120]. Stabilization of G4 at replication forks can lead to significant genomic instability and DNA breakage.

6.2. G4 Unwinders

G4 destabilizers, also known as G4 destabilizers, work by promoting the dissociation of G4 structures to prevent replication fork stalling and DNA breakage. For example, Pif1 DNA helicase and 2′-F cytidine trimers (2′-F C3). G4 structures are formed in vivo, and they are resolved by Pif1 DNA helicase [112]. 2′-F C3 can release the translation of mRNA containing G-quadruplexes without inducing DNA damage [24]. In the context of immunotherapy, G4 unwinders can be used to modify the expression of cytokines, immune checkpoint genes, and other immune-regulatory factors, thereby enhancing anti-tumor immune responses. The use of G4 unwinders in combination with immune checkpoint inhibitors may enhance the efficacy of immunotherapies by promoting immune cell activation and cytokine production.

6.3. Combination Therapies

Given the complex role of G4 in regulating immune checkpoint genes and tumor growth, combination therapy targeting G4 structures combined with traditional immunotherapy and chemotherapy drugs has significant prospects. G4-targeted drugs can be used in combination with immune checkpoint inhibitors (such as anti-PD-1/PD-L1 antibodies) to form a double-hit strategy, that is, reducing PD-L1 expression through G4 regulation and enhancing T cell function through PD-1 inhibitors to improve the therapeutic effect [16]. Chemotherapeutic drugs that induce DNA damage (such as cisplatin) can synergize with G4-targeted therapy to further induce genomic instability in tumor cells, thereby enhancing the overall therapeutic effect [113]. In addition, G4 regulation may also enhance the effect of tumor vaccines or CAR-T therapy, but the current dose matching, treatment window, and long-term safety still need further exploration [114].

6.4. Multifunctional Nanomedicines

Nanotechnology has shown great potential in G4-targeted therapy. For example, nanocarriers for targeted delivery of G4 stabilizers can improve drug selectivity, stability, and bioavailability while reducing systemic toxicity [115,116]. In addition, nanomaterials can also be used to co-deliver G4-targeted drugs with immunotherapeutic molecules, such as nanoencapsulated anti-PD-L1 antibodies combined with G4 modulators to enhance anti-tumor immune responses. Multifunctional nanomedicines that combine G4 modulators with immune checkpoint inhibitors, chemotherapeutic drugs, or cytokines have the potential to enhance the therapeutic effects of cancer immunotherapy [117]. For example, nanoparticles encapsulating G4 stabilizers can be designed to selectively release drugs at tumor sites, improve pharmacokinetics, and minimize off-target effects These nanocarriers can also be used to co-deliver anti-PD-L1 antibodies to enhance anti-tumor immune responses while modulating G4-related immune pathways.

7. Conclusions and Outlook

In summary, G4 structures are emerging as critical regulatory elements that influence gene transcription, DNA replication, telomere maintenance, and RNA metabolism. Recent evidence underscores their pivotal role in modulating the tumor immune microenvironment by affecting immune checkpoint expression, cytokine production, immune cell function, and various molecular interaction networks. These multifaceted functions position G4 as a promising target for innovative cancer immunotherapeutic strategies.

Despite the rapid advances in G-quadruplex (G4)-targeted cancer therapy, several critical challenges must be addressed before clinical translation can be achieved. G4-stabilizing ligands, such as CX-3543 (Quarfloxin) and CX-5461, have shown promising antitumor activity by stabilizing G4 structures at oncogene promoters or rDNA regions, leading to transcriptional repression and DNA damage [16,121–123]. However, both compounds exhibit off-target effects and limited pharmacokinetic profiles; notably, CX-3543 failed in clinical trials due to poor bioavailability and high plasma protein binding [123]. Moreover, the dynamic nature of G4 structures across different genomic contexts complicates their selective targeting in cancer versus normal tissues [124]. Targeting G4-unwinding helicases, such as WRN, has gained attention, especially in microsatellite instability-high (MSI-H) tumors where WRN is synthetically essential [125]. WRN inhibitors like NSC 617145 and ML216 sensitize cells to replication stress, yet their systemic inhibition may also impair genome stability in normal proliferating cells [126], raising concerns of potential toxicity. Combination therapies present a rational strategy to enhance the efficacy of G4-targeted agents. For instance, CX-5461 synergizes with PARP inhibitors in BRCA-deficient tumors by exacerbating replication stress [127], and G4 ligands have also been shown to boost immune checkpoint blockade by enhancing tumor immunogenicity type I interferon responses [4,18].

Nevertheless, combination regimens must be cautiously optimized to avoid overlapping toxicities and unintended immunosuppressive effects. To improve tumor-specific delivery and reduce systemic exposure, various nanoparticle-based systems have been developed to encapsulate G4 ligands. For example, liposomal formulations of GQC-05, a G4-stabilizing ligand, improved its solubility and tumor targeting in preclinical leukemia models [128]. Similarly, AS1411, a G-quadruplex-forming aptamer, was utilized as a tumor-targeted delivery vehicle for TMPyP4. The AS1411–TMPyP4 complex exhibited enhanced cellular uptake and accumulation in tumor cells, leading to improved photodynamic therapeutic efficacy [129]. However, challenges such as immune clearance, off-target uptake, tumor heterogeneity, and large-scale production still limit the clinical application of these systems. An additional concern is the emergence of resistance mechanisms. Tumors may adapt to G4 stress by upregulating G4-resolving helicases (e.g., BLM, PIF1), mutating G4-forming regions, or reprogramming DNA repair pathways [130,131]. To overcome such resistance, researchers are exploring dual-function compounds that simultaneously target G4s and helicases, or combining G4 ligands with immune modulators to reinforce therapeutic pressure. Importantly, the lack of robust in vivo models that faithfully recapitulate the immunological and transcriptional consequences of G4-targeting remains a major limitation. The development of G4-reporter mouse models, immune-competent tumor-bearing mice, and spatial transcriptomic tools to visualize G4 dynamics will be essential for future mechanistic and therapeutic validation.

In summary, while G4-targeted strategies hold immense promise for tumor immunomodulation, their translation requires solving multiple pharmacological, biological, and immunological challenges. Addressing these issues through innovative drug design, biomarker-driven patient selection, and mechanistically guided combination regimens will be crucial for unlocking the full potential of G4 biology in cancer therapy.

Looking ahead, the integration of advanced multi-omics analyses with nanotechnology is expected to accelerate the identification of novel G4 biomarkers and therapeutic targets. This interdisciplinary approach could facilitate the development of personalized cancer immunotherapies that leverage the unique regulatory capabilities of G4 structures. Future research should prioritize the refinement of G4-targeting agents for enhanced specificity and minimal side effects, as well as the design of rigorous clinical trials to evaluate their safety and efficacy.

Ultimately, continued efforts to understand and manipulate G4-mediated regulatory networks hold promise for overcoming current limitations in cancer immunotherapy and may establish G4-targeted strategies as a cornerstone of precision oncology.

Author Contributions: H.Z. and Y.Z. (Yunxia Zhou) drafted the manuscript, and Y.Z. (Ying Zhang) and D.X. critically reviewed and commented on the manuscript. All authors have read and agreed to the published version of the manuscript.

Funding: This work was supported by the Shenzhen Science, Technology, and Innovation Commission (SZSTI) Basic Research Program (JCYJ20210324120601005; JCYJ20190809165417340); the National Natural Science Foundation of China (31900429), and the Natural Science Foundation of Guangdong Province, China (2021A1515011070; 2019A1515110812).

Institutional Review Board Statement: Not applicable.

Informed Consent Statement: Not applicable.

Conflicts of Interest: The authors declare no conflicts of interest.

References

1. Bochman, M.L.; Paeschke, K.; Zakian, V.A. DNA secondary structures: Stability and function of G-quadruplex structures. *Nat. Rev. Genet.* **2012**, *13*, 770–780. [CrossRef] [PubMed]
2. Rhodes, D.; Lipps, H.J. G-quadruplexes and their regulatory roles in biology. *Nucleic Acids Res.* **2015**, *43*, 8627–8637. [CrossRef]
3. Johnson, F.B. Fundamentals of G-quadruplex biology. *Annu. Rep. Med. Chem.* **2020**, *54*, 3–44. [CrossRef]
4. Li, P.; Zhou, D.; Xie, Y.; Yuan, Z.; Huang, M.; Xu, G.; Huang, J.; Zhuang, Z.; Luo, Y.; Yu, H.; et al. Targeting G-quadruplex by TMPyP4 for inhibition of colorectal cancer through cell cycle arrest and boosting anti-tumor immunity. *Cell Death Dis.* **2024**, *15*, 816. [CrossRef]
5. Vijay Kumar, M.J.; Mitteaux, J.; Wang, Z.; Wheeler, E.; Tandon, N.; Yun Jung, S.; Hudson, R.H.E.; Monchaud, D.; Tsvetkov, A.S. Small molecule-based regulation of gene expression in human astrocytes switching on and off the G-quadruplex control systems. *J. Biol. Chem.* **2025**, *301*, 108040. [CrossRef]
6. Pavlova, A.V.; Savitskaya, V.Y.; Dolinnaya, N.G.; Monakhova, M.V.; Litvinova, A.V.; Kubareva, E.A.; Zvereva, M.I. G-Quadruplex Formed by the Promoter Region of the hTERT Gene: Structure-Driven Effects on DNA Mismatch Repair Functions. *Biomedicines* **2022**, *10*, 1871. [CrossRef] [PubMed]
7. Shen, Z.; Zheng, R.; Yang, H.; Xing, S.; Jin, X.; Yan, H.; Zhu, J.; Mei, Y.; Lin, F.; Zheng, X. G-quadruplex stabilizer Tetra-Pt(bpy) disrupts telomere maintenance and impairs FAK-mediated migration of telomerase-positive cells. *Int. J. Biol. Macromol.* **2022**, *213*, 858–870. [CrossRef] [PubMed]
8. Gellert, M.; Lipsett, M.N.; Davies, D.R. Helix formation by guanylic acid. *Proc. Natl. Acad. Sci. USA* **1962**, *48*, 2013–2018. [CrossRef]
9. Wang, X.D.; Lin, J.H.; Hu, M.H. Discovery of a tribenzophenazine analog for binding to the KRAS mRNA G-quadruplex structures in the cisplatin-resistant non-small cell lung cancer. *J. Biol. Chem.* **2025**, *301*, 108164. [CrossRef]
10. Zheng, B.X.; Long, W.; Zheng, W.; Zeng, Y.; Guo, X.C.; Chan, K.H.; She, M.T.; Leung, A.S.; Lu, Y.J.; Wong, W.L. Mitochondria-Selective Dicationic Small-Molecule Ligand Targeting G-Quadruplex Structures for Human Colorectal Cancer Therapy. *J. Med. Chem.* **2024**, *67*, 6292–6312. [CrossRef]
11. Taylor, J.P. Neurodegenerative diseases: G-quadruplex poses quadruple threat. *Nature* **2014**, *507*, 175–177. [CrossRef]
12. Yan, M.P.; Wee, C.E.; Yen, K.P.; Stevens, A.; Wai, L.K. G-quadruplex ligands as therapeutic agents against cancer, neurological disorders and viral infections. *Future Med. Chem.* **2023**, *15*, 1987–2009. [CrossRef]
13. Kaulage, M.H.; Maji, B.; Pasadi, S.; Ali, A.; Bhattacharya, S.; Muniyappa, K. Targeting G-quadruplex DNA structures in the telomere and oncogene promoter regions by benzimidazole—carbazole ligands. *Eur. J. Med. Chem.* **2018**, *148*, 178–194. [CrossRef] [PubMed]
14. Sarkar, S.; Bisoi, A.; Singh, P.C. Antimalarial Drugs Induce the Selective Folding of Human Telomeric G-Quadruplex in a Cancer-Mimicking Microenvironment. *J. Phys. Chem. B* **2023**, *127*, 6648–6655. [CrossRef] [PubMed]
15. Shankar, U.; Jain, N.; Mishra, S.K.; Sharma, T.K.; Kumar, A. Conserved G-Quadruplex Motifs in Gene Promoter Region Reveals a Novel Therapeutic Approach to Target Multi-Drug Resistance Klebsiella pneumoniae. *Front. Microbiol.* **2020**, *11*, 1269. [CrossRef]
16. Chung, S.Y.; Chang, Y.C.; Hsu, D.S.; Hung, Y.C.; Lu, M.L.; Hung, Y.P.; Chiang, N.J.; Yeh, C.N.; Hsiao, M.; Soong, J.; et al. A G-quadruplex stabilizer, CX-5461 combined with two immune checkpoint inhibitors enhances in vivo therapeutic efficacy by increasing PD-L1 expression in colorectal cancer. *Neoplasia* **2023**, *35*, 100856. [CrossRef]
17. Yao, L.; Wang, L.; Liu, S.; Qu, H.; Mao, Y.; Li, Y.; Zheng, L. Evolution of a bispecific G-quadruplex-forming circular aptamer to block IL-6/sIL-6R interaction for inflammation inhibition. *Chem. Sci.* **2024**, *15*, 13011–13020. [CrossRef] [PubMed]
18. Miglietta, G.; Russo, M.; Duardo, R.C.; Capranico, G. G-quadruplex binders as cytostatic modulators of innate immune genes in cancer cells. *Nucleic Acids Res.* **2021**, *49*, 6673–6686. [CrossRef]
19. Shi, K.; Chen, K.; Lu, S.; Luo, Q.; Xu, Q. G-quadruplex structures in FGFR3 promoter negatively regulate its gene expression and DNA replication. *Biochem. Biophys. Res. Commun.* **2024**, *730*, 150384. [CrossRef]
20. Zacheja, T.; Toth, A.; Harami, G.M.; Yang, Q.; Schwindt, E.; Kovacs, M.; Paeschke, K.; Burkovics, P. Mgs1 protein supports genome stability via recognition of G-quadruplex DNA structures. *FASEB J.* **2020**, *34*, 12646–12662. [CrossRef]
21. Ye, H.; Zhang, H.; Xiang, J.; Shen, G.; Yang, F.; Wang, F.; Wang, J.; Tang, Y. Advances and prospects of natural dietary polyphenols as G-quadruplex stabilizers in biomedical applications. *Int. J. Biol. Macromol.* **2024**, *254*, 127825. [CrossRef]
22. Zegers, J.; Peters, M.; Albada, B. DNA G-quadruplex-stabilizing metal complexes as anticancer drugs. *J. Biol. Inorg. Chem.* **2023**, *28*, 117–138. [CrossRef]
23. Liu, L.Y.; Ma, T.Z.; Zeng, Y.L.; Liu, W.; Zhang, H.; Mao, Z.W. Organic-Platinum Hybrids for Covalent Binding of G-Quadruplexes: Structural Basis and Application to Cancer Immunotherapy. *Angew. Chem. Int. Ed. Engl.* **2023**, *62*, e202305645. [CrossRef]

24. Teng, X.; Hu, D.; Dai, Y.; Jing, H.; Hu, W.; Zhang, Q.; Zhang, N.; Li, J. Discovery of A G-Quadruplex Unwinder That Unleashes the Translation of G-Quadruplex-Containing mRNA without Inducing DNA Damage. *Angew. Chem. Int. Ed. Engl.* **2024**, *63*, e202407353. [CrossRef] [PubMed]
25. Seneff, S.; Nigh, G.; Kyriakopoulos, A.M.; McCullough, P.A. Innate immune suppression by SARS-CoV-2 mRNA vaccinations: The role of G-quadruplexes, exosomes, and MicroRNAs. *Food Chem. Toxicol.* **2022**, *164*, 113008. [CrossRef] [PubMed]
26. Teng, F.Y.; Jiang, Z.Z.; Guo, M.; Tan, X.Z.; Chen, F.; Xi, X.G.; Xu, Y. G-quadruplex DNA: A novel target for drug design. *Cell. Mol. Life Sci.* **2021**, *78*, 6557–6583. [CrossRef] [PubMed]
27. Lv, L.; Zhang, L. Characterization of G-Quadruplexes in Enterovirus A71 Genome and Their Interaction with G-Quadruplex Ligands. *Microbiol. Spectr.* **2022**, *10*, e0046022. [CrossRef]
28. Matsuo, K.; Asamitsu, S.; Maeda, K.; Suzuki, H.; Kawakubo, K.; Komiya, G.; Kudo, K.; Sakai, Y.; Hori, K.; Ikenoshita, S.; et al. RNA G-quadruplexes form scaffolds that promote neuropathological alpha-synuclein aggregation. *Cell* **2024**, *187*, 6835–6848.e20. [CrossRef]
29. Pirota, V.; Stritto, A.D.; Magnaghi, L.R.; Biesuz, R.; Doria, F.; Mella, M.; Freccero, M.; Crespan, E. A Novel G-Quadruplex Structure within Apolipoprotein E Promoter: A New Promising Target in Cancer and Dementia Fight? *ACS Omega* **2024**, *9*, 45203–45213. [CrossRef]
30. Wang, X.D.; Wang, J.X.; Hu, M.H. Novel phenanthrene imidazoles as telomeric G-quadruplex ligands trigger potent immunogenic cell death in triple-negative breast cancer. *Int. J. Biol. Macromol.* **2023**, *249*, 126068. [CrossRef]
31. Chan, K.H.; Zheng, B.X.; Leung, A.S.; Long, W.; Zhao, Y.; Zheng, Y.; Wong, W.L. A NRAS mRNA G-quadruplex structure-targeting small-molecule ligand reactivating DNA damage response in human cancer cells for combination therapy with clinical PI3K inhibitors. *Int. J. Biol. Macromol.* **2024**, *279*, 135308. [CrossRef]
32. Balasubramanian, S.; Hurley, L.H.; Neidle, S. Targeting G-quadruplexes in gene promoters: A novel anticancer strategy? *Nat. Rev. Drug Discov.* **2011**, *10*, 261–275. [CrossRef]
33. Kosiol, N.; Juranek, S.; Brossart, P.; Heine, A.; Paeschke, K. G-quadruplexes: A promising target for cancer therapy. *Mol. Cancer* **2021**, *20*, 40. [CrossRef]
34. Varshney, D.; Spiegel, J.; Zyner, K.; Tannahill, D.; Balasubramanian, S. The regulation and functions of DNA and RNA G-quadruplexes. *Nat. Rev. Mol. Cell Biol.* **2020**, *21*, 459–474. [CrossRef] [PubMed]
35. Wu, Y.L.; Horwitz, N.E.; Chen, K.S.; Gomez-Gualdron, D.A.; Luu, N.S.; Ma, L.; Wang, T.C.; Hersam, M.C.; Hupp, J.T.; Farha, O.K.; et al. G-quadruplex organic frameworks. *Nat. Chem.* **2017**, *9*, 466–472. [CrossRef] [PubMed]
36. Liu, S.; Peng, P.; Wang, H.; Shi, L.; Li, T. Thioflavin T binds dimeric parallel-stranded GA-containing non-G-quadruplex DNAs: A general approach to lighting up double-stranded scaffolds. *Nucleic Acids Res.* **2017**, *45*, 12080–12089. [CrossRef] [PubMed]
37. Chen, L.; Dickerhoff, J.; Sakai, S.; Yang, D. DNA G-Quadruplex in Human Telomeres and Oncogene Promoters: Structures, Functions, and Small Molecule Targeting. *Acc. Chem. Res.* **2022**, *55*, 2628–2646. [CrossRef]
38. Pavc, D.; Wang, B.; Spindler, L.; Drevensek-Olenik, I.; Plavec, J.; Sket, P. GC ends control topology of DNA G-quadruplexes and their cation-dependent assembly. *Nucleic Acids Res.* **2020**, *48*, 2749–2761. [CrossRef]
39. Liu, S.G.; Luo, D.; Han, L.; Li, N.B.; Luo, H.Q. A hybrid material composed of guanine-rich single stranded DNA and cobalt(III) oxyhydroxide (CoOOH) nanosheets as a fluorescent probe for ascorbic acid via formation of a complex between G-quadruplex and thioflavin T. *Mikrochim. Acta* **2019**, *186*, 156. [CrossRef]
40. Boncina, M.; Vesnaver, G.; Chaires, J.B.; Lah, J. Unraveling the Thermodynamics of the Folding and Interconversion of Human Telomere G-Quadruplexes. *Angew. Chem. Int. Ed. Engl.* **2016**, *55*, 10340–10344. [CrossRef]
41. Beseiso, D.; Chen, E.V.; McCarthy, S.E.; Martin, K.N.; Gallagher, E.P.; Miao, J.; Yatsunyk, L.A. The first crystal structures of hybrid and parallel four-tetrad intramolecular G-quadruplexes. *Nucleic Acids Res.* **2022**, *50*, 2959–2972. [CrossRef]
42. Biver, T. Discriminating between Parallel, Anti-Parallel and Hybrid G-Quadruplexes: Mechanistic Details on Their Binding to Small Molecules. *Molecules* **2022**, *27*, 4165. [CrossRef] [PubMed]
43. El-Khoury, R.; Roman, M.; Assi, H.A.; Moye, A.L.; Bryan, T.M.; Damha, M.J. Telomeric i-motifs and C-strands inhibit parallel G-quadruplex extension by telomerase. *Nucleic Acids Res.* **2023**, *51*, 10395–10410. [CrossRef]
44. Sergeev, A.V.; Loiko, A.G.; Genatullina, A.I.; Petrov, A.S.; Kubareva, E.A.; Dolinnaya, N.G.; Gromova, E.S. Crosstalk between G-Quadruplexes and Dnmt3a-Mediated Methylation of the c-MYC Oncogene Promoter. *Int. J. Mol. Sci.* **2023**, *25*, 45. [CrossRef] [PubMed]
45. D'Aria, F.; Pagano, B.; Petraccone, L.; Giancola, C. KRAS Promoter G-Quadruplexes from Sequences of Different Length: A Physicochemical Study. *Int. J. Mol. Sci.* **2021**, *22*, 448. [CrossRef]
46. Pandya, N.; Singh, M.; Rani, R.; Kumar, V.; Kumar, A. G-quadruplex-mediated specific recognition, stabilization and transcriptional repression of bcl-2 by small molecule. *Arch. Biochem. Biophys.* **2023**, *734*, 109483. [CrossRef] [PubMed]
47. Lyu, J.; Shao, R.; Kwong Yung, P.Y.; Elsasser, S.J. Genome-wide mapping of G-quadruplex structures with CUT&Tag. *Nucleic Acids Res.* **2022**, *50*, e13. [CrossRef]
48. Hegyi, H. Enhancer-promoter interaction facilitated by transiently forming G-quadruplexes. *Sci. Rep.* **2015**, *5*, 9165. [CrossRef]

49. Zhang, R.; Wang, Y.; Wang, C.; Sun, X.; Mergny, J.L. G-quadruplexes as pivotal components of cis-regulatory elements in the human genome. *BMC Biol.* **2024**, *22*, 177. [CrossRef]
50. Poulet-Benedetti, J.; Tonnerre-Doncarli, C.; Valton, A.L.; Laurent, M.; Gerard, M.; Barinova, N.; Parisis, N.; Massip, F.; Picard, F.; Prioleau, M.N. Dimeric G-quadruplex motifs-induced NFRs determine strong replication origins in vertebrates. *Nat. Commun.* **2023**, *14*, 4843. [CrossRef]
51. Park, D.; Chung, W.C.; Gong, S.; Ravichandran, S.; Lee, G.M.; Han, M.; Kim, K.K.; Ahn, J.H. G-quadruplex as an essential structural element in cytomegalovirus replication origin. *Nat. Commun.* **2024**, *15*, 7353. [CrossRef] [PubMed]
52. Mou, X.; Liew, S.W.; Kwok, C.K. Identification and targeting of G-quadruplex structures in MALAT1 long non-coding RNA. *Nucleic Acids Res.* **2022**, *50*, 397–410. [CrossRef]
53. Luige, J.; Armaos, A.; Tartaglia, G.G.; Orom, U.A.V. Predicting nuclear G-quadruplex RNA-binding proteins with roles in transcription and phase separation. *Nat. Commun.* **2024**, *15*, 2585. [CrossRef] [PubMed]
54. Shukla, C.; Datta, B. G-quadruplexes in long non-coding RNAs and their interactions with proteins. *Int. J. Biol. Macromol.* **2024**, *278*, 134946. [CrossRef]
55. Ghafouri-Fard, S.; Abak, A.; Baniahmad, A.; Hussen, B.M.; Taheri, M.; Jamali, E.; Dinger, M.E. Interaction between non-coding RNAs, mRNAs and G-quadruplexes. *Cancer Cell Int.* **2022**, *22*, 171. [CrossRef]
56. Liu, G.; Du, W.; Xu, H.; Sun, Q.; Tang, D.; Zou, S.; Zhang, Y.; Ma, M.; Zhang, G.; Du, X.; et al. RNA G-quadruplex regulates microRNA-26a biogenesis and function. *J. Hepatol.* **2020**, *73*, 371–382. [CrossRef] [PubMed]
57. Li, F.; Guo, D.; Xie, T.; Zhang, S.; Wang, A.; Li, Y.; Zhou, J. G-quadruplex from precursor miR-1587 modulated its maturation and function. *Int. J. Biol. Macromol.* **2023**, *231*, 123279. [CrossRef] [PubMed]
58. Chambers, V.S.; Marsico, G.; Boutell, J.M.; Di Antonio, M.; Smith, G.P.; Balasubramanian, S. High-throughput sequencing of DNA G-quadruplex structures in the human genome. *Nat. Biotechnol.* **2015**, *33*, 877–881. [CrossRef]
59. Kwok, C.K.; Marsico, G.; Sahakyan, A.B.; Chambers, V.S.; Balasubramanian, S. rG4-seq reveals widespread formation of G-quadruplex structures in the human transcriptome. *Nat. Methods* **2016**, *13*, 841–844. [CrossRef]
60. Galli, S.; Flint, G.; Ruzickova, L.; Di Antonio, M. Genome-wide mapping of G-quadruplex DNA: A step-by-step guide to select the most effective method. *RSC Chem. Biol.* **2024**, *5*, 426–438. [CrossRef]
61. Siddiqui-Jain, A.; Grand, C.L.; Bearss, D.J.; Hurley, L.H. Direct evidence for a G-quadruplex in a promoter region and its targeting with a small molecule to repress c-MYC transcription. *Proc. Natl. Acad. Sci. USA* **2002**, *99*, 11593–11598. [CrossRef] [PubMed]
62. Lin, C.; Yang, D. Human Telomeric G-Quadruplex Structures and G-Quadruplex-Interactive Compounds. *Methods Mol. Biol.* **2017**, *1587*, 171–196. [CrossRef]
63. Wolfe, A.L.; Singh, K.; Zhong, Y.; Drewe, P.; Rajasekhar, V.K.; Sanghvi, V.R.; Mavrakis, K.J.; Jiang, M.; Roderick, J.E.; Van der Meulen, J.; et al. RNA G-quadruplexes cause eIF4A-dependent oncogene translation in cancer. *Nature* **2014**, *513*, 65–70. [CrossRef] [PubMed]
64. Marsico, G.; Chambers, V.S.; Sahakyan, A.B.; McCauley, P.; Boutell, J.M.; Antonio, M.D.; Balasubramanian, S. Whole genome experimental maps of DNA G-quadruplexes in multiple species. *Nucleic Acids Res.* **2019**, *47*, 3862–3874. [CrossRef]
65. Wu, W.; Rokutanda, N.; Takeuchi, J.; Lai, Y.; Maruyama, R.; Togashi, Y.; Nishikawa, H.; Arai, N.; Miyoshi, Y.; Suzuki, N.; et al. HERC2 Facilitates BLM and WRN Helicase Complex Interaction with RPA to Suppress G-Quadruplex DNA. *Cancer Res.* **2018**, *78*, 6371–6385. [CrossRef]
66. Dai, Y.; Teng, X.; Zhang, Q.; Hou, H.; Li, J. Advances and challenges in identifying and characterizing G-quadruplex-protein interactions. *Trends Biochem. Sci.* **2023**, *48*, 894–909. [CrossRef] [PubMed]
67. Qin, G.; Liu, Z.; Yang, J.; Liao, X.; Zhao, C.; Ren, J.; Qu, X. Targeting specific DNA G-quadruplexes with CRISPR-guided G-quadruplex-binding proteins and ligands. *Nat. Cell Biol.* **2024**, *26*, 1212–1224. [CrossRef]
68. Yuan, J.; He, X.; Wang, Y. G-quadruplex DNA contributes to RNA polymerase II-mediated 3D chromatin architecture. *Nucleic Acids Res.* **2023**, *51*, 8434–8446. [CrossRef]
69. Ma, T.Z.; Zhang, M.J.; Liao, T.C.; Li, J.H.; Zou, M.; Wang, Z.M.; Zhou, C.Q. Dimers formed with the mixed-type G-quadruplex binder pyridostatin specifically recognize human telomere G-quadruplex dimers. *Org. Biomol. Chem.* **2020**, *18*, 920–930. [CrossRef]
70. Kharel, P.; Ivanov, P. RNA G-quadruplexes and stress: Emerging mechanisms and functions. *Trends Cell Biol.* **2024**, *34*, 771–784. [CrossRef]
71. Cammas, A.; Dubrac, A.; Morel, B.; Lamaa, A.; Touriol, C.; Teulade-Fichou, M.P.; Prats, H.; Millevoi, S. Stabilization of the G-quadruplex at the VEGF IRES represses cap-independent translation. *RNA Biol.* **2015**, *12*, 320–329. [CrossRef] [PubMed]
72. Sun, J.W.; Zou, J.; Zheng, Y.; Yuan, H.; Xie, Y.Z.; Wang, X.N.; Ou, T.M. Design, synthesis, and evaluation of novel quindoline derivatives with fork-shaped side chains as RNA G-quadruplex stabilizers for repressing oncogene NRAS translation. *Eur. J. Med. Chem.* **2024**, *271*, 116406. [CrossRef] [PubMed]
73. Chen, S.B.; Hu, M.H.; Liu, G.C.; Wang, J.; Ou, T.M.; Gu, L.Q.; Huang, Z.S.; Tan, J.H. Visualization of NRAS RNA G-Quadruplex Structures in Cells with an Engineered Fluorogenic Hybridization Probe. *J. Am. Chem. Soc.* **2016**, *138*, 10382–10385. [CrossRef]

74. Gao, J.; Pickett, H.A. Targeting telomeres: Advances in telomere maintenance mechanism-specific cancer therapies. *Nat. Rev. Cancer* **2022**, *22*, 515–532. [CrossRef]
75. Lansdorp, P.M. Telomeres, aging, and cancer: The big picture. *Blood* **2022**, *139*, 813–821. [CrossRef]
76. Elango, H.; Das, R.N.; Saha, A. Benzimidazole-based small molecules as anticancer agents targeting telomeric G-quadruplex and inhibiting telomerase enzyme. *Future Med. Chem.* **2024**, *16*, 2043–2067. [CrossRef] [PubMed]
77. Stavnezer, J.; Guikema, J.E.; Schrader, C.E. Mechanism and regulation of class switch recombination. *Annu. Rev. Immunol.* **2008**, *26*, 261–292. [CrossRef]
78. Tang, C.; MacCarthy, T. Characterization of DNA G-Quadruplex Structures in Human Immunoglobulin Heavy Variable (IGHV) Genes. *Front. Immunol.* **2021**, *12*, 671944. [CrossRef]
79. Deze, O.; Laffleur, B.; Cogne, M. Roles of G4-DNA and G4-RNA in Class Switch Recombination and Additional Regulations in B-Lymphocytes. *Molecules* **2023**, *28*, 1159. [CrossRef]
80. Dharmaiah, S.; Malgulwar, P.B.; Johnson, W.E.; Chen, B.A.; Sharin, V.; Whitfield, B.T.; Alvarez, C.; Tadimeti, V.; Farooqi, A.S.; Huse, J.T. G-quadruplex stabilizer CX-5461 effectively combines with radiotherapy to target ATRX-deficient malignant glioma. *Neuro-Oncology* **2024**, noae248. [CrossRef]
81. Groelly, F.J.; Porru, M.; Zimmer, J.; Benainous, H.; De Visser, Y.; Kosova, A.A.; Di Vito, S.; Serra, V.; Ryan, A.; Leonetti, C.; et al. Anti-tumoural activity of the G-quadruplex ligand pyridostatin against BRCA1/2-deficient tumours. *EMBO Mol. Med.* **2022**, *14*, e14501. [CrossRef]
82. Escarcega, R.D.; Patil, A.A.; Moruno-Manchon, J.F.; Urayama, A.; Marrelli, S.P.; Kim, N.; Monchaud, D.; McCullough, L.D.; Tsvetkov, A.S. Pirh2-dependent DNA damage in neurons induced by the G-quadruplex ligand pyridostatin. *J. Biol. Chem.* **2023**, *299*, 105157. [CrossRef]
83. Kallweit, L.; Hamlett, E.D.; Saternos, H.; Gilmore, A.; Granholm, A.C.; Horowitz, S. Chronic RNA G-quadruplex Accumulation in Aging and Alzheimer's Disease. *eLife* **2025**, *14*, RP105446. [CrossRef] [PubMed]
84. Riccardi, C.; D'Aria, F.; Fasano, D.; Digilio, F.A.; Carillo, M.R.; Amato, J.; De Rosa, L.; Paladino, S.; Melone, M.A.B.; Montesarchio, D.; et al. Truncated Analogues of a G-Quadruplex-Forming Aptamer Targeting Mutant Huntingtin: Shorter Is Better! *Int. J. Mol. Sci.* **2022**, *23*, 12412. [CrossRef]
85. Blice-Baum, A.C.; Mihailescu, M.R. Biophysical characterization of G-quadruplex forming FMR1 mRNA and of its interactions with different fragile X mental retardation protein isoforms. *RNA* **2014**, *20*, 103–114. [CrossRef]
86. Amrane, S.; Jaubert, C.; Bedrat, A.; Rundstadler, T.; Recordon-Pinson, P.; Aknin, C.; Guedin, A.; De Rache, A.; Bartolucci, L.; Diene, I.; et al. Deciphering RNA G-quadruplex function during the early steps of HIV-1 infection. *Nucleic Acids Res.* **2022**, *50*, 12328–12343. [CrossRef] [PubMed]
87. Perrone, R.; Butovskaya, E.; Lago, S.; Garzino-Demo, A.; Pannecouque, C.; Palu, G.; Richter, S.N. The G-quadruplex-forming aptamer AS1411 potently inhibits HIV-1 attachment to the host cell. *Int. J. Antimicrob. Agents* **2016**, *47*, 311–316. [CrossRef] [PubMed]
88. Qin, G.; Zhao, C.; Liu, Y.; Zhang, C.; Yang, G.; Yang, J.; Wang, Z.; Wang, C.; Tu, C.; Guo, Z.; et al. RNA G-quadruplex formed in SARS-CoV-2 used for COVID-19 treatment in animal models. *Cell Discov.* **2022**, *8*, 86. [CrossRef]
89. Xi, H.; Juhas, M.; Zhang, Y. G-quadruplex based biosensor: A potential tool for SARS-CoV-2 detection. *Biosens. Bioelectron.* **2020**, *167*, 112494. [CrossRef]
90. Ma, T.Z.; Liu, L.Y.; Zeng, Y.L.; Ding, K.; Zhang, H.; Liu, W.; Cao, Q.; Xia, W.; Xiong, X.; Wu, C.; et al. G-quadruplex-guided cisplatin triggers multiple pathways in targeted chemotherapy and immunotherapy. *Chem. Sci.* **2024**, *15*, 9756–9774. [CrossRef]
91. Leppek, K.; Das, R.; Barna, M. Functional 5′ UTR mRNA structures in eukaryotic translation regulation and how to find them. *Nat. Rev. Mol. Cell Biol.* **2018**, *19*, 158–174. [CrossRef] [PubMed]
92. Tellam, J.T.; Zhong, J.; Lekieffre, L.; Bhat, P.; Martinez, M.; Croft, N.P.; Kaplan, W.; Tellam, R.L.; Khanna, R. mRNA Structural constraints on EBNA1 synthesis impact on in vivo antigen presentation and early priming of CD8+ T cells. *PLoS Pathog.* **2014**, *10*, e1004423. [CrossRef]
93. Biswas, S.; Samui, S.; Das, A.K.; Pasadi, S.; Muniyappa, K.; Naskar, J. Targeting G-quadruplex DNA with synthetic dendritic peptide: Modulation of the proliferation of human cancer cells. *RSC Adv.* **2020**, *10*, 26388–26396. [CrossRef]
94. Liao, W.; Tan, M.; Kusamori, K.; Takakura, Y.; Nishikawa, M. Construction of Monomeric and Dimeric G-Quadruplex-Structured CpG Oligodeoxynucleotides for Enhanced Uptake and Activation in TLR9-Positive Macrophages. *Nucleic Acid. Ther.* **2020**, *30*, 299–311. [CrossRef] [PubMed]
95. Wang, J.X.; Wang, X.D.; Hu, M.H. Novel quinoxaline analogs as telomeric G-quadruplex ligands exert antitumor effects related to enhanced immunomodulation. *Eur. J. Med. Chem.* **2024**, *274*, 116536. [CrossRef] [PubMed]
96. Wang, X.D.; Liu, Y.S.; Chen, M.D.; Hu, M.H. Discovery of a triphenylamine-based ligand that targets mitochondrial DNA G-quadruplexes and activates the cGAS-STING immunomodulatory pathway. *Eur. J. Med. Chem.* **2024**, *269*, 116361. [CrossRef]
97. Matsumoto, S.; Tateishi-Karimata, H.; Sugimoto, N. DNA methylation is regulated by both the stability and topology of G-quadruplex. *Chem. Commun.* **2022**, *58*, 12459–12462. [CrossRef]

98. Song, J.; Gooding, A.R.; Hemphill, W.O.; Love, B.D.; Robertson, A.; Yao, L.; Zon, L.I.; North, T.E.; Kasinath, V.; Cech, T.R. Structural basis for inactivation of PRC2 by G-quadruplex RNA. *Science* **2023**, *381*, 1331–1337. [CrossRef]
99. Tassinari, M.; Richter, S.N.; Gandellini, P. Biological relevance and therapeutic potential of G-quadruplex structures in the human noncoding transcriptome. *Nucleic Acids Res.* **2021**, *49*, 3617–3633. [CrossRef]
100. Kleemann, J.; Steinhorst, K.; Konig, V.; Zoller, N.; Cinatl, J., Jr.; Ozistanbullu, D.; Kaufmann, R.; Meissner, M.; Kippenberger, S. Functional Downregulation of PD-L1 and PD-L2 by CpG and non-CpG Oligonucleotides in Melanoma Cells. *Cancers* **2022**, *14*, 4698. [CrossRef]
101. Propper, D.J.; Balkwill, F.R. Harnessing cytokines and chemokines for cancer therapy. *Nat. Rev. Clin. Oncol.* **2022**, *19*, 237–253. [CrossRef] [PubMed]
102. Niu, K.; Zhang, X.; Song, Q.; Feng, Q. G-Quadruplex Regulation of VEGFA mRNA Translation by RBM4. *Int. J. Mol. Sci.* **2022**, *23*, 743. [CrossRef]
103. Wang, X.; Chen, S.; Zhao, Z.; Chen, F.; Huang, Y.; Guo, X.; Lei, L.; Wang, W.; Luo, Y.; Yu, H.; et al. Genomic G-quadruplex folding triggers a cytokine-mediated inflammatory feedback loop to aggravate inflammatory diseases. *iScience* **2022**, *25*, 105312. [CrossRef] [PubMed]
104. Wang, H.; Wang, L.; Guo, S.; Liu, Z.; Zhao, L.; Qiao, R.; Li, C. Rutin-Loaded Stimuli-Responsive Hydrogel for Anti-Inflammation. *ACS Appl. Mater. Interfaces* **2022**, *14*, 26327–26337. [CrossRef] [PubMed]
105. Kuryavyi, V.; Phan, A.T.; Patel, D.J. Solution structures of all parallel-stranded monomeric and dimeric G-quadruplex scaffolds of the human c-kit2 promoter. *Nucleic Acids Res.* **2010**, *38*, 6757–6773. [CrossRef]
106. Gunaratnam, M.; Collie, G.W.; Reszka, A.P.; Todd, A.K.; Parkinson, G.N.; Neidle, S. A naphthalene diimide G-quadruplex ligand inhibits cell growth and down-regulates BCL-2 expression in an imatinib-resistant gastrointestinal cancer cell line. *Bioorg. Med. Chem.* **2018**, *26*, 2958–2964. [CrossRef]
107. Wang, X.; Zhou, C.X.; Yan, J.W.; Hou, J.Q.; Chen, S.B.; Ou, T.M.; Gu, L.Q.; Huang, Z.S.; Tan, J.H. Synthesis and Evaluation of Quinazolone Derivatives as a New Class of c-KIT G-Quadruplex Binding Ligands. *ACS Med. Chem. Lett.* **2013**, *4*, 909–914. [CrossRef]
108. Shankar, U.; Mishra, S.K.; Jain, N.; Tawani, A.; Yadav, P.; Kumar, A. Ni^{+2} permease system of Helicobacter pylori contains highly conserved G-quadruplex motifs. *Infect. Genet. Evol.* **2022**, *101*, 105298. [CrossRef]
109. Xu, H.; Hurley, L.H. A first-in-class clinical G-quadruplex-targeting drug. The bench-to-bedside translation of the fluoroquinolone QQ58 to CX-5461 (Pidnarulex). *Bioorg. Med. Chem. Lett.* **2022**, *77*, 129016. [CrossRef]
110. Liu, L.Y.; Ma, T.Z.; Zeng, Y.L.; Liu, W.; Mao, Z.W. Structural Basis of Pyridostatin and Its Derivatives Specifically Binding to G-Quadruplexes. *J. Am. Chem. Soc.* **2022**, *144*, 11878–11887. [CrossRef]
111. Shitikov, E.; Bespiatykh, D.; Malakhova, M.; Bespyatykh, J.; Bodoev, I.; Vedekhina, T.; Zaychikova, M.; Veselovsky, V.; Klimina, K.; Ilina, E.; et al. Genome-Wide Transcriptional Response of Mycobacterium smegmatis MC(2)155 to G-Quadruplex Ligands BRACO-19 and TMPyP4. *Front. Microbiol.* **2022**, *13*, 817024. [CrossRef] [PubMed]
112. Paeschke, K.; Capra, J.A.; Zakian, V.A. DNA replication through G-quadruplex motifs is promoted by the Saccharomyces cerevisiae Pif1 DNA helicase. *Cell* **2011**, *145*, 678–691. [CrossRef]
113. Kuang, K.; Li, C.; Maksut, F.; Ghosh, D.; Vinck, R.; Wang, M.; Poupon, J.; Xiang, R.; Li, W.; Li, F.; et al. A G-quadruplex-binding platinum complex induces cancer mitochondrial dysfunction through dual-targeting mitochondrial and nuclear G4 enriched genome. *J. Biomed. Sci.* **2024**, *31*, 50. [CrossRef] [PubMed]
114. Ruden, M.; Puri, N. Novel anticancer therapeutics targeting telomerase. *Cancer Treat. Rev.* **2013**, *39*, 444–456. [CrossRef]
115. Dong, J.; O'Hagan, M.P.; Willner, I. Switchable and dynamic G-quadruplexes and their applications. *Chem. Soc. Rev.* **2022**, *51*, 7631–7661. [CrossRef] [PubMed]
116. Mao, X.; Zhang, X.; Chao, Z.; Qiu, D.; Wei, S.; Luo, R.; Chen, D.; Zhang, Y.; Chen, Y.; Yang, Y.; et al. A Versatile G-Quadruplex (G4)-Coated Upconverted Metal-Organic Framework for Hypoxic Tumor Therapy. *Adv. Healthc. Mater.* **2023**, *12*, e2300561. [CrossRef]
117. Clua, A.; Fabrega, C.; Garcia-Chica, J.; Grijalvo, S.; Eritja, R. Parallel G-quadruplex Structures Increase Cellular Uptake and Cytotoxicity of 5-Fluoro-2′-deoxyuridine Oligomers in 5-Fluorouracil Resistant Cells. *Molecules* **2021**, *26*, 1741. [CrossRef]
118. Liu, L.; Geng, X.; Zhang, J.; Li, S.; Gao, J. Structure-based discovery of Licoflavone B and Ginkgetin targeting c-Myc G-quadruplex to suppress c-Myc transcription and myeloma growth. *Chem. Biol. Drug Des.* **2022**, *100*, 525–533. [CrossRef]
119. Jin, M.; Hurley, L.H.; Xu, H. A synthetic lethal approach to drug targeting of G-quadruplexes based on CX-5461. *Bioorg. Med. Chem. Lett.* **2023**, *91*, 129384. [CrossRef]
120. McMullen, M.; Karakasis, K.; Madariaga, A.; Oza, A.M. Overcoming Platinum and PARP-Inhibitor Resistance in Ovarian Cancer. *Cancers* **2020**, *12*, 1607. [CrossRef]
121. Lee, H.C.; Wang, H.; Baladandayuthapani, V.; Lin, H.; He, J.; Jones, R.J.; Kuiatse, I.; Gu, D.; Wang, Z.; Ma, W.; et al. RNA Polymerase I Inhibition with CX-5461 as a Novel Therapeutic Strategy to Target MYC in Multiple Myeloma. *Br. J. Haematol.* **2017**, *177*, 80–94. [CrossRef] [PubMed]

122. Xu, H.; Di Antonio, M.; McKinney, S.; Mathew, V.; Ho, B.; O'Neil, N.J.; Santos, N.D.; Silvester, J.; Wei, V.; Garcia, J.; et al. CX-5461 is a DNA G-quadruplex stabilizer with selective lethality in BRCA1/2 deficient tumours. *Nat. Commun.* **2017**, *8*, 14432. [CrossRef] [PubMed]
123. Drygin, D.; Siddiqui-Jain, A.; O'Brien, S.; Schwaebe, M.; Lin, A.; Bliesath, J.; Ho, C.B.; Proffitt, C.; Trent, K.; Whitten, J.P.; et al. Anticancer activity of CX-3543: A direct inhibitor of rRNA biogenesis. *Cancer Res.* **2009**, *69*, 7653–7661. [CrossRef] [PubMed]
124. Biffi, G.; Di Antonio, M.; Tannahill, D.; Balasubramanian, S. Visualization and selective chemical targeting of RNA G-quadruplex structures in the cytoplasm of human cells. *Nat. Chem.* **2014**, *6*, 75–80. [CrossRef]
125. Chan, E.M.; Shibue, T.; McFarland, J.M.; Gaeta, B.; Ghandi, M.; Dumont, N.; Gonzalez, A.; McPartlan, J.S.; Li, T.; Zhang, Y.; et al. WRN helicase is a synthetic lethal target in microsatellite unstable cancers. *Nature* **2019**, *568*, 551–556. [CrossRef]
126. Moles, R.; Bai, X.T.; Chaib-Mezrag, H.; Nicot, C. WRN-targeted therapy using inhibitors NSC 19630 and NSC 617145 induce apoptosis in HTLV-1-transformed adult T-cell leukemia cells. *J. Hematol. Oncol.* **2016**, *9*, 121. [CrossRef]
127. Sanij, E.; Hannan, K.M.; Xuan, J.; Yan, S.; Ahern, J.E.; Trigos, A.S.; Brajanovski, N.; Son, J.; Chan, K.T.; Kondrashova, O.; et al. CX-5461 activates the DNA damage response and demonstrates therapeutic efficacy in high-grade serous ovarian cancer. *Nat. Commun.* **2020**, *11*, 2641. [CrossRef]
128. Montoya, J.J.; Turnidge, M.A.; Wai, D.H.; Patel, A.R.; Lee, D.W.; Gokhale, V.; Hurley, L.H.; Arceci, R.J.; Wetmore, C.; Azorsa, D.O. In vitro activity of a G-quadruplex-stabilizing small molecule that synergizes with Navitoclax to induce cytotoxicity in acute myeloid leukemia cells. *BMC Cancer* **2019**, *19*, 1251. [CrossRef]
129. Shieh, Y.A.; Yang, S.J.; Wei, M.F.; Shieh, M.J. Aptamer-based tumor-targeted drug delivery for photodynamic therapy. *ACS Nano* **2010**, *4*, 1433–1442. [CrossRef]
130. Mendoza, O.; Bourdoncle, A.; Boule, J.B.; Brosh, R.M., Jr.; Mergny, J.L. G-quadruplexes and helicases. *Nucleic Acids Res.* **2016**, *44*, 1989–2006. [CrossRef]
131. Wang, Y.R.; Guo, T.T.; Zheng, Y.T.; Lai, C.W.; Sun, B.; Xi, X.G.; Hou, X.M. Replication protein A plays multifaceted roles complementary to specialized helicases in processing G-quadruplex DNA. *iScience* **2021**, *24*, 102493. [CrossRef] [PubMed]

biomedicines MDPI

Article

The IL-6/JAK/STAT3 Axis in Cholangiocarcinoma and Primary Sclerosing Cholangitis: Unlocking Therapeutic Strategies Through Patient-Derived Organoids

Corinna Boden [1], Laura K. Esser [2], Leona Dold [3], Bettina Langhans [3], Taotao Zhou [3], Dominik J. Kaczmarek [3], Maria A. Gonzalez-Carmona [3], Tobias J. Weismüller [4], Glen Kristiansen [2], Jörg C. Kalff [1], Michael Hölzel [5], Hanno Matthaei [1], Marieta I. Toma [2] and Vittorio Branchi [1,*]

1 Department of General, Abdominal, Thoracic and Vascular Surgery, University Hospital Bonn, 53127 Bonn, Germany; corinna.boden@ukbonn.de (C.B.); joerg.kalff@ukbonn.de (J.C.K.); hanno.matthaei@ukbonn.de (H.M.)
2 Institute of Pathology, University Hospital Bonn, 53127 Bonn, Germany; glen.kristiansen@ukbonn.de (G.K.)
3 Department of Internal Medicine I, University Hospital Bonn, 53127 Bonn, Germany; leona.dold@ukbonn.de (L.D.); bettina.langhans@ukbonn.de (B.L.); taotao.zhou@ukbonn.de (T.Z.); dominik.kaczmarek@ukbonn.de (D.J.K.); maria.gonzalez-carmona@ukbonn.de (M.A.G.-C.)
4 Vivantes Humboldt-Klinikum, 13509 Berlin, Germany; tobias.weismueller@vivantes.de
5 Institute of Experimental Oncology, University Hospital Bonn, 53127 Bonn, Germany; michael.hoelzel@ukbonn.de
* Correspondence: vittorio.branchi@ukbonn.de

Abstract: Background/Objectives: Primary sclerosing cholangitis (PSC) is a rare, incurable liver disease characterized by chronic biliary inflammation and fibrosis. PSC is a significant risk factor for biliary tract cancer (BTC). This study aims to evaluate STAT3 expression in BTC and its prognostic significance as well as explore the potential of organoids derived from PSC and liver tumor patients as an in vitro model for testing novel therapeutic strategies in both PSC and BTC. **Methods:** Fresh tissue samples obtained from 10 PSC patients through targeted endoscopic retrograde cholangiography (ERC) and biopsy samples from liver tumor patients were used to establish organoid cultures. Organoids were treated with different agents and the therapeutic effect was measured by CellTiterGlo. Treatment with the JAK inhibitor baricitinib was followed by the measurement of cytokine concentrations in the supernatant. Archived formalin-fixed paraffin-embedded (FFPE) samples from 55 surgically resected BTC tumors were analyzed for STAT3 expression using immunohistochemistry. **Results:** We successfully established organoid cultures from all ERC samples. STAT3 protein expression was detected in 56% of tumor samples and 69% of the immune microenvironment. STAT3 positivity in the immune cell compartment was associated with longer disease-free survival, although the multivariate analysis could not confirm its value as an independent prognostic factor. Chemotherapy testing on liver tumor organoids showed various degrees of decreases in viability after treatment with gemcitabine, cisplatin, and cabozantinib. Baricitinib treatment significantly reduced IL-6 and MCP-1 secretion in cholangiocarcinoma **Conclusions:** The patient-derived organoid model of PSC and liver tumors is a valuable tool for testing novel and established therapeutic strategies, including JAK inhibitors and chemotherapy regimens. STAT3 expression in the immune microenvironment of BTC may serve as a prognostic marker. Further studies are needed to explore the integration of co-cultured organoid systems with stromal and immune components to improve physiological relevance.

Keywords: biliary tract cancer; cholangiocarcinoma; primary sclerosing cholangitis; STAT3; baricitinib

1. Introduction

Primary sclerosing cholangitis (PSC) is a rare, incurable, cholestatic liver disease characterized by chronic, progressive biliary inflammation and subsequent fibrosis [1]. The etiopathogenesis of PSC remains unclear and effective therapy is lacking [2]. The incidence of PSC varies in different parts of the world and is estimated to range from 0 to 1.58 per 100,000, whereas the median prevalence ranges between 0 and 24.99 per 100,000 [3]. Men have almost twice the risk of developing PSC as women [4]. PSC is found to co-exist with inflammatory bowel disease (IBD) in 60–80% of cases, with a higher prevalence of IBD in male PSC patients compared to females [5]. On the other hand, PSC occurs in approximately 8% of individuals with IBD, and the most common IBD diagnosed in PSC patients is ulcerative colitis [6]. PSC is considered a premalignant condition and represents a risk factor for various abdominal malignancies, which account for 40–50% of all deaths in patients with PSC [7–10]. Biliary tract cancer (BTC), in particular cholangiocarcinoma (CCA), is the most common malignancy and the leading cause of death in patients with PSC [11]. Early detection of BTC in PSC remains a major clinical challenge, and 1-year mortality reaches 80% in patients diagnosed with PSC-associated BTC [11]. Endoscopic retrograde cholangiography (ERC) is a crucial tool for the diagnostic evaluation and treatment of high-grade strictures in PSC. However, its use should be reserved for cases with unclear diagnoses or when tissue sampling or endoscopic interventions are necessary [12]. The limited understanding of PSC pathogenesis and the scarcity of effective therapeutic options highlight the critical need for translational research in this area.

Biliary tract cancer (BTC) refers to a heterogeneous group of aggressive, malignant tumors that arise along the bile ducts (cholangiocarcinoma—CCA) and in the gallbladder or cystic duct (gallbladder carcinoma—GBC) [13]. Most individuals with BTC receive a diagnosis at an advanced, unresectable stage, primarily because symptoms are typically absent during the initial phases of the disease [14]. BTCs are further classified as intrahepatic CCA (iCCA), which arises from bile ducts proximal to the second-order bile ducts; perihilar CCA (phCCA), which arises at the junction of the right and left hepatic ducts or within the right or left hepatic duct; distal CCA (dCCA), which arises distal to the cystic duct insertion [15]. In most countries, CCA remains a relatively rare malignancy; however, its incidence and mortality rates have steadily risen over the past few decades [16]. Incidence and mortality rate, as well as risk factors for CCA, vary substantially based on geographic regions [17]. In region where liver fluke is endemic, such as Thailand, CCA age-standardized incidence can reach 115 per 100,000 in men and 49 per 100,000 in women [18]. In regions where liver fluke is not considered endemic, the CCA incidence ranges between 0.3 and 6 per 100,000 [19]. Among the various risk factors for CCA—such as cirrhosis, cholelithiasis/choledocholithiasis, hepatitis, liver flukes, bile duct cysts, and Caroli's disease—primary sclerosing cholangitis (PSC) stands out as the most significant risk factor in western countries [17,20]. GBC is a rare tumor with a global age-standardized incidence rate of 0.9 for males and 1.4 for females [21]. Risk factors for GBC include cholelithiasis, obesity, and bile duct infections [22].

Signal transducer and activator of transcription 3 (STAT3) is a transcription factor that belongs to the STAT family and plays a role in normal cellular processes, such as cell survival and proliferation [23]. In cancer, STAT3 overexpression has been observed in many solid tumors, as well as in hematologic malignancies, correlating with unfavorable clinical stages, and the worst overall and disease-free survival [24–27]. Hyperactivation of the STAT3 pathway contributes to key cancer hallmarks, including epithelial-to-mesenchymal transition (EMT), angiogenesis, and metastasis [28]. STAT3 signaling is initiated when cytokines or growth factors (e.g., IL-6, IL-10, EGF, PDGF, TNF) bind to their receptors, activating receptor-associated Janus kinases (JAKs). Activated JAKs phosphorylate each other and the receptor's tyrosine residues, creating docking sites for STAT3. STAT3 is recruited and phosphorylated by JAKs, enabling its dimerization. The STAT3 dimer translocates to the nucleus, binds to specific DNA sequences, and activates transcription of a broad number of genes, including cancer promoting genes [29,30]. STAT3 has been identified as a key gene involved in all three main human cholangiopathies, including PSC [31]. In BTC, STAT3 overexpression has been associated with worse pathological characteristics, as well as negative surgical outcomes, and some data identified STAT3 as a driver of cancer proliferation and metastasis [32,33].

This study aimed to evaluate STAT3 expression in BTC and assess its prognostic significance in a cohort of BTC patients. In addition, it aimed to establish a biobank of organoids derived from both PSC and liver tumor patients to explore their potential as an in vitro model for testing novel therapeutic strategies for PSC and BTC. In this study, we aimed to use cholangioscopy-guided biopsies of the main strictures of PSC patients to provide a highly specific and minimally invasive system for organoid generation. Biopsies taken directly from the site of the main stricture may provide a targeted and clinically relevant approach for PSC organoid generation.

2. Materials and Methods

2.1. FFPE Samples from Surgically Resected Tumors

We included archived samples from patients who underwent surgical resection for BTC at the Department of Surgery. Patients' demographic data, including gender, age, and tumor and treatment-related data, were collected. Survival data and pathological characteristics were retrieved from patients' records. All tumors were thoroughly restaged according to the TNM classification, 8th Edition.

2.2. Immunohistochemistry

Antibodies to STAT3 and CK7 were used. Briefly, 3 µm sections from FFPE blocks were mounted on Tomo® Adhesion microscope slides, (Matsunami Glass Ind. LTD, Osaka, Japan). The sections were deparaffinized with xylene (2 × 15 min, VWR International, Fontenay-sous-Bois, France) and rehydrated using decreasing concentrations of graded ethanol (Berkel AHK, Ludwigshafen, Germany) to water (B. Braun, Melsungen, Germany). Antigen retrieval was achieved by boiling the slides in citrate buffer (Zytomed Systems GmbH, Berlin, Germany) at pH 6.0 for 20 min. The tissue samples were then stained overnight at 4 °C. Immunohistochemistry was performed using the semi-automated platform Autostainer 480 S (Medac, Wedel, Germany). All supplementary reagents were provided by Medac.

2.3. Analysis of Immunoreactivity

A Tissue microarray (TMA) was constructed according to standardized protocols from FFPE blocks. First, standard hematoxylin–eosin (H&E) staining of 3 µmsection was obtained to select tumor regions. Four to six 1 mm cores per sample were transferred to the TMA. STAT3 protein expression, both in immune cells and in tumor cells, was evaluated. An expert pathologist double-checked the assessment of immune reactivity. In the immune cells compartment, the presence of a membranous or a cytoplasmic staining was considered a positive immunoreactivity. A semi-quantitative scoring system was applied, where samples were classified as STAT3-positive if more than 10% of immune cells showed positivity for STAT3. In tumor cells, STAT3 expression was evaluated as the percentage of positive cells. Mean values were calculated for each tumor. Tumor samples with less than 10% positive tumor cells were considered STAT3-negative.

2.4. Fresh Tissue Samples from PSC Patients and from Patients with Liver Malignancies

Biopsies from patients with PSC who underwent an elective endoscopic retrograde cholagiography (ERC) were obtained. The indication for ERC was made according to clinical evaluation and independently from this study. The ERC was performed according to clinical standards by highly experienced endoscopists. Cholangioscopies were performed with single-use, single-operator-controlled devices (SpyScopeTM DS, Boston Scientific, Marlborough, IL, USA). Biopsies for routine pathological examinations were made according to clinical evaluations. Two extra samples were obtained at the site of pathological strictures with a single-use, biopsy forceps with a 0.8 mm-wide oval spoon-shaped mouth with tooth (Micro Byte, MTW-Endoskopie, Wesel, Germany) or with a standard 2.3 mm biopsy forceps with oval cups with spikes (Endo-Flex GmbH, Voerde, Germany). The samples were directly placed in a 15 mL falcon tube filled with ice-cold Dulbecco's PBS after collection. Within 30 min after collection, the samples were processed to obtain the extrahepatic cholangiocyte organoids (ECOs). Biopsy from surgery specimens were obtained from patients with liver tumors who underwent liver resection. The indication for surgical resection was made independently from this study through an interdisciplinary tumor board. Within 30 min of collection, the samples were then processed to obtain patient-derived organoids. Every patient included in this study signed the informed consent prior to the investigation/operation according to our institutional recommendations and in compliance with the Declaration of Helsinki.

2.5. Culture of Extrahepatic Cholangiocyte Organoid from PSC Patients and Tumor Organoids Derived from Primary or Metastatic Tumors of the Liver

Patient-derived organoids were cultured using an adaptation of previously described methods [34]. Fresh tissue biopsies and biopsy-like samples from surgical specimens were washed in wash media (Supplementary Table S1), minced, and incubated at 37 °C with a digestion solution (Supplementary Table S2). After 30 min of incubation, the suspension was washed three times with the wash media, filtered through a 100 μm nylon cell strainer, and spun for 5 min at 300 G. The pellet was washed in cold Dulbecco's PBS, then mixed with Geltrex matrix (Gibco, Thermofisher, Wahltam, MA, USA) and plated as domes in a pre-warmed 6-multi-well plate. After the 15 min incubation (37 °C, 5% CO_2), 1.5 mL of organoid isolation media (Supplementary Table S4) was added into each well. Organoid cultures were incubated at 37 °C with 5% CO_2, and media were replaced twice a week. After one to two weeks, depending on the growth pattern, organoid isolation media were replaced with organoid expansion media (Supplementary Table S5), or organoid tumoroid media for HCC organoids (Supplementary Table S6). Once they reached a considerable density or dimension, organoids were passaged by mechanical dissociation into small fragments, transferred to fresh Geltrex matrix, and replated as domes in pre-warmed 6-multi-well plates. A mycoplasma PCR test was performed every 4 weeks. Control organoids were generated from bile duct epithelial tissue obtained from a non-diseased gallbladder, which served as external controls in the cytokine secretion experiments.

2.6. Organoid Treatment with a JAK Inhibitor and Measurement of Cytokines in the Organoid Supernatant

Cholangiocyte organoids were treated with a reversible Janus kinase JAK1 and JAK2 inhibitor (baricitinib). First, organoids were cultivated at least 1 week before starting the treatment. Organoids were then passaged and seeded in a 96-well plate (Sarstedt AG & Co. KG, Nümbrecht, Germany). After 72 h of incubation, organoids were treated with baricitinib at different concentrations (1 nM, 10 nM, and 100 nM). The supernatant was then collected and centrifuged to remove debris. Cytokine concentrations in supernatants were then measured through flow cytometry using the LEGENDplex HU

Essential Immune Response Panel (Biolegend, San Diego, CA, USA). Samples were analyzed on a FACSCanto II (BD Biosciences, Heidelberg, Germany). Data evaluation was performed using LEGENDplex data analysis software (Biolegend, San Diego, CA, USA, https://www.biolegend.com/fr-fr/immunoassays/legendplex, accessed on 15 January 2025). In total, 13 different cytokines were analyzed: IL-4, IL-2, CCL10 (IP-10), IL-1β, TNF-α, TGF-β1, CXCL8 (IL-8), CCL2 (MCP-1), IL-17A, IL-6, IL-10, IFN-γ, and IL-12.

2.7. DNA Sequencing

Whole genomic DNA was extracted from cultured organoids using the innuCONVERT bisulfite all-in-one kit (Analytik Jena, Jena, Germany). Multiplex amplicon preparation and subsequent library construction were performed using the TruSight Tumor 15 Panel (Illumina, San Diego, CA, USA). The amplification products were deep-sequenced on the MiSeq (Illumina, San Diego, CA, USA) with a sequence coverage of >500 reads. VariantStudio 3.0 (Illumina, San Diego, CA, USA) was used for data analysis and identification of genomic variants. All relevant somatic mutations with an allele frequency >5% were reported. The current versions of the following online databases were used for classification and reporting of somatic variants: dbSNP, ClinVar, OncoKB, and ExAc.

2.8. Chemotherapy Testing

Patient-derived organoids from malignant primary liver tumors (N = 3) and liver metastasis (N = 2) were cultivated for 1 week before starting the treatment. The organoids were then passaged and seeded in a 96-well plate (Sarstedt, Germany). After 72h, organoids were treated with gemcitabine, cisplatin, or cabozantinib, either alone or in combinations, at different concentrations. Gemcitabine was used at the following concentrations: 5 μM, 10 μM, 20 μM, and 40 μM. Cisplatin working concentrations were 2.5 μM, 5 μM, 10 μM, and 20 μM. Cabozantinib was used at 4.5 μM, 9 μM, 18 μM, and 36 μM. A combination of gemcitabine and cisplatin (5/2.5 μM, 10/5 μM, 20/10 μM, 40/20 μM) was also used. After 72 h, 100-μL CellTiter-Glo®-Luminescent reagent (Promega, Madison, WI, US) per well was added and incubated for 20 min. The medium was then transferred to a nontransparent 96-well plate (Greiner Bio-One GmbH, Frickenhausen, Austria) and luminescence was assessed on a Centro LB 960 microplate luminometer (Berthold Technologies GmbH & Co. KG, Bad Wildbad, Germany). Every experiment was carried out as a triplicate.

2.9. Statistical Analysis

Statistical analysis was performed in the R environment (RStudio Version 12, packages: ggplot, survival, survminer, gt, and gtsummary) [35–41] and Graphpad Prism (Version 10, GraphPad Software, Boston, MA, USA). The Wilcoxon rank sum test, Pearson's chi-squared test, and Fisher's exact test were used as appropriate. A log-rank test was used to compare Kaplan–Meier curves. A p value < 0.05 was considered significant.

3. Results

3.1. Establishment of a Patient-Derived Organoid Biobank for PSC

Biopsies from ERC were obtained from ten consecutive patients, five females (50%) and five males (50%). The median age at PSC diagnosis was 37 (interquartile range 23–47). The median age of the patients at the time of the sampling was 45 years (interquartile range 39–49). In two cases, the samples were obtained at the patient's first endoscopic retrograde cholangiography (ERC). Patient characteristics are summarized in Table 1.

Table 1. Patients' characteristics of the ERC-organoid cohort.

Sample Number	Age at ERCP	Sex	Age at First Diagnosis	Sampling	Amsterdam Score	Dominant Stricture	Dysplasia	Cancer	Diagnosis	Pathology Notes
1	64	F	49	HDB	3	HDB	0	0	PSC	/
2	49	M	17	LHD	3	LHD	0	0	PSC	increased focal IG4 Plasma cells
3	49	M	48	LHD	2	LHD	0	0	PSC	/
4	34	F	34	HDB	2	HDB	0	0	PSC	/
5	47	F	33	LHD	3	LHD	0	0	PSC	/
6	27	M	21	LHD	3	LHD	0	0	PSC	/
7	42	M	42	HDB	3	HDB	0	0	PSC	/
8	41	M	23	LHD	3	LHD	0	0	PSC	/
9	49	F	47	HDB	/	HDB	0	0	Benign stricture	/
checked10	39	F	39	CBD	2	CBD	0	0	PSC	/

HDB: hepatic duct bifurcation; LHD: left hepatic duct; CBD: common bile duct; PSC: primary sclerosing cholangitis.

First, the main stricture was identified during the cholangioscopy. After the routine biopsies for clinical use were performed, two additional biopsy specimens from a previously identified main stricture were obtained for this study. An organoid culture could be successfully accomplished in 10 out of the 10 (100%) cases. The organoids were passaged regularly for a median of six times (range: 3–13, median: 93 days, range: 36–158) before being frozen. Every organoid was stored at −196 °C and thawed at least twice. After the thawing process, all organoids remained vital for at least 15 days. Every organoid line could be passaged for the first time after 8–20 days (Figure 1D). All organoids from different patients appeared phenotypically identical and displayed positivity for CK7 (Figure 1G–H, Supplementary Table S7). In one case, a sample was replated on day one in culture because of too-dense cellularity. The left hepatic duct (LHD) was sampled most frequently (n = 5, 50%), followed by hepatic duct bifurcation (HDB, n = 4, 40%). In one case, the biopsies were obtained from the common bile duct (CBD), just below the cystic duct junction. In one case, the PSC diagnosis could not be definitively confirmed, and the patient was lost to follow-up. In two out of ten samples, we could identify deleterious mutations (Supplementary Table S8).

3.2. Establishment of a Patient-Derived Organoid Biobank from Liver Tumors

In parallel to the samples from PSC patients, we collected biopsy-like samples from patients undergoing surgery for liver tumors. A total of 14 organoid lines were successfully established from eight patients with primary cholangiocarcinoma, two patients with hepatocellular carcinoma (HCC), and three patients with liver metastasis from colorectal malignancies (CRCm). The patients' characteristics and site of sampling are summarized in Table 2.

The success rate of establishing a long-term organoid line (≥10 passages) was 60% for CRCm (3/5), 57% for cholangiocarcinoma (8/14), and 33% for HCC (2/6). Every organoid line could be passaged after 7–22 days, stored at −196 °C, and thawed at least twice.

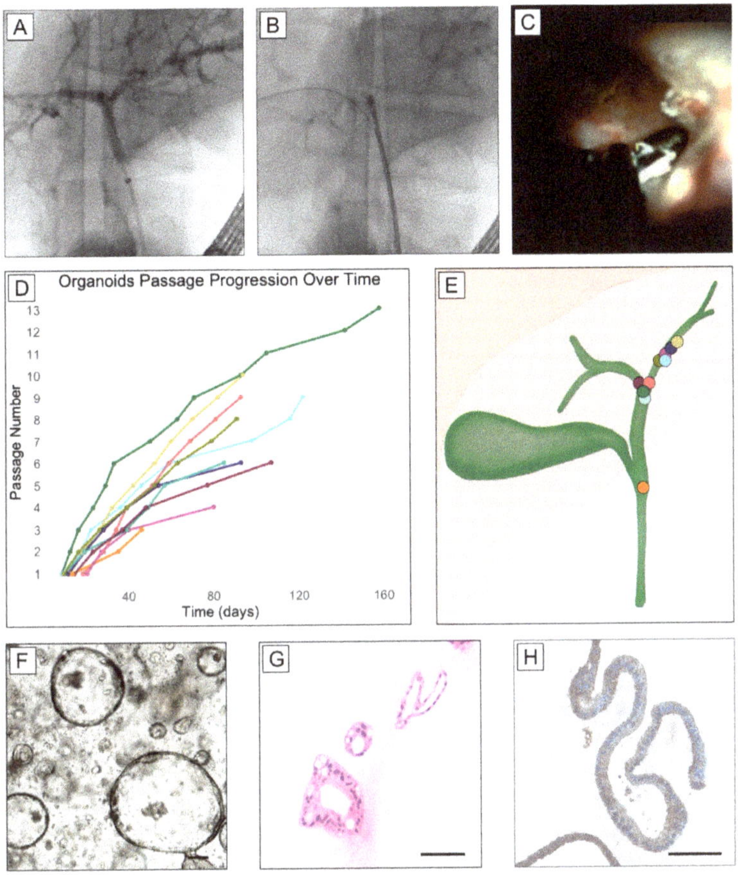

Figure 1. Exemplary images of an endoscopic retrograde cholangiography (ERC) of a main stricture site at the hepatic duct bifurcation after a balloon dilatation (**A**) and cholangiography-guided biopsy at the main stricture site (**B**), as well as a cholangioscopy-guided biopsy of a main stricture (**C**). Timeline of ERC-organoid passages (**D**). Graphical representation of ERC sampling sites to obtain ERC-derived organoids. The organoids show similar growth pattern regardless of the sampling site (**E**). Representative images of ERC-derived organoids: bright-field live microscopy image (**F**), hematoxylin–eosin staining (**G**), and immunohistochemistry staining (**H**) for cytokeratin-7 after formalin fixation and paraffin embedding. Scale bar = 100 µm.

3.3. A Patient-Derived Organoid Model Can Be Used to Test Chemotherapy Regimens in Different Liver Malignancies

To determine whether the organoid model responds to chemotherapy in a way that suggests its usability for real-life testing, we performed an exploratory chemotherapy test on PDOs from different liver tumors using different agents. In one CCA organoid, a decrease in viability was observed after 48 h of treatment with gemcitabine and the combination of gemcitabine and cisplatin. However, cisplatin alone did not affect viability, while cabozantinib treatment resulted in a moderate viability reduction. The patient was given oral capecitabine. After 15 months of follow-up, he had no signs of relapse (Figure 2A). In one HCC organoid, a similar decrease in viability was observed after treatment with gemcitabine alone and in combination with gemcitabine and cisplatin. Some effects were also observed after treatment with cisplatin and cabozantinib at higher concentrations (Figure 2B). The patient was enrolled in a trial and did not receive any chemotherapy. After

12 months of follow-up, no signs of relapse were observed. In the colorectal organoid, a decrease in viability was observed after treatment with gemcitabine and cisplatin, alone or in combination, and after treatment with cabozantinib. The patient did not receive any adjuvant treatment and was disease-free after 24 months of follow-up (Figure 2C).

Table 2. Patient characteristics of the tumor-derived organoid cohort.

Age	Sex	Tumor Type	Sampling Site	T	N	M	G	L	V	Pn	R
61	M	phCCA	Segment III	pT2b	pN1	M0	G2	L1	V0	Pn0	R0
72	F	iCCA	Segment V	pT3	pN0	M0	G2	L0	V1	Pn1	R1
68	M	phCCA	Segment IVa	pT2b	pN0	M0	G2	L0	V0	Pn1	R1
82	F	HCC	Segment III	pT2	pNx	M0	G3	L0	V0	Pn0	R0
73	F	iCCA	Segment IVa	1b	pN0	M0	G2	L0	V0	Pn0	R1
76	M	CRC Met	Segment III	pT4a	pN0	M1	G2	L0	V0	Pn0	R0
78	M	HCC	Segment VIII	pT2	pN0	M0	G3	L0	V1	Pn0	R0
75	M	dCCA	CBD	pT3	pN0	M0	G2	L0	V0	Pn0	R0
75	M	HCC	Segment II	pT2	pNx	M0	G2	L0	V1	Pn0	R0
42	F	CRC Met	Segment V	ypT2	ypN0	M2	NA	L0	V0	Pn0	R0
80	F	phCCA	Segment III	pT1a	pN0	M0	G1	L0	V0	Pn0	R0
73	M	CRC Met	Segment V	pT2	pN0	M2	NA	L0	V0	Pn0	R0
60	M	HCC	Segment IVb	pT3	pN0	M0	G2	L0	V1	Pn0	R0

phCCA: perihilar cholangiocarcinoma; iCCA: intrahepatic cholangiocarcinoma; HCC: hepatocellular carcinoma; dCCA: distal cholangiocarcinoma; CRC Met: colorectal cancer metastasis; CBD: common bile duct.

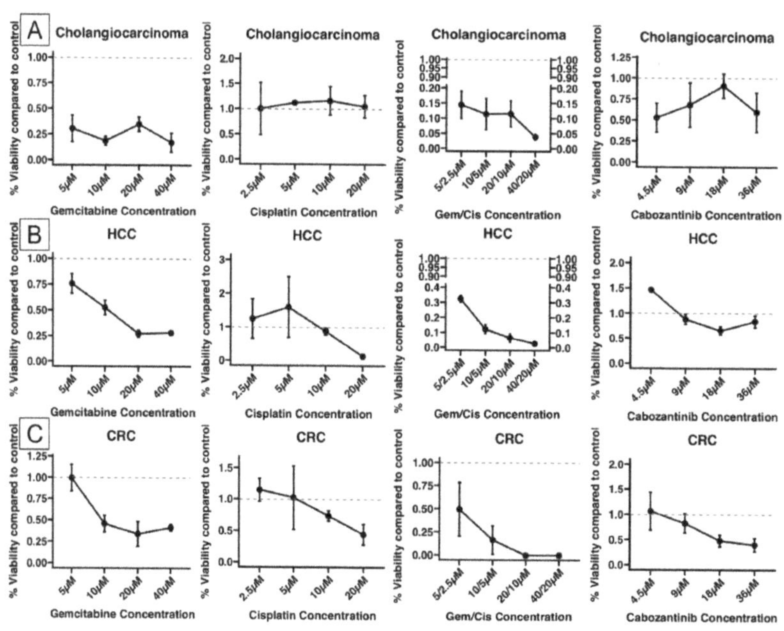

Figure 2. Viability after 48 h of chemotherapy testing on organoids from (**A**) cholangiocarcinoma, (**B**) hepatocellular carcinoma (HCC), and (**C**) colorectal metastasis (CRC). The mean viability and standard error of the mean are displayed. Viability is normalized to the control (Gem/Cis: gemcitabine/cisplatin).

3.4. Baricitinib Reduces IL-6 and MCP1 Secretion in Cholangiocarcinoma and May Have an Effect on PSC Cholangiocytes' Secretion

To assess the potential of the organoid model for PSC and cholangiocarcinoma in novel therapy testing, we investigated the effect of baricitinib, a JAK inhibitor, on cytokine secretion. In cholangiocarcinoma, a significant reduction in IL-6 secretion was observed in the supernatant after 48 h of incubation (log concentration 3.37 vs. 2.32, $p = 0.038$). The concentration of MCP1 was also significantly reduced after 48h of treatment (log concentration: 3.36 vs. 2.31, $p = 0.034$) (Figure 3). However, baricitinib had no effect on other cytokines, as IP-10 and IL-8 concentrations remained stable after 48 h of treatment. In the PSC organoid model, a slight decrease in IL-8 and IP-10 concentrations was detected following baricitinib treatment, though this reduction was not statistically significant. The IL-6 concentration in the supernatant of ERC-derived organoids was below the lowest measurable concentration. No significant alterations were observed in the secretion of other cytokines after treatment.

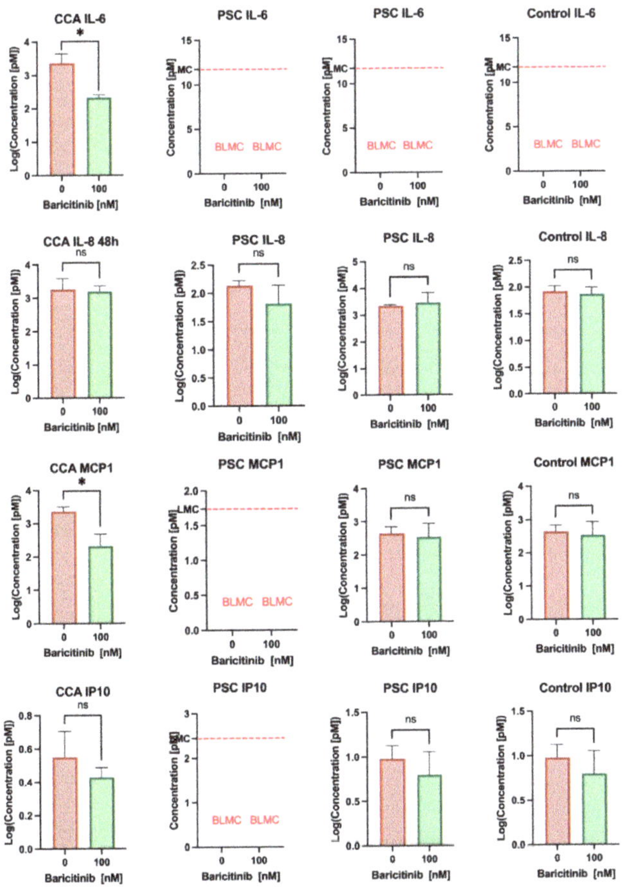

Figure 3. Bar plots displaying the log (concentration) of various cytokines after 48 h of treatment with baricitinib. Resulkts from one CCA organoid, two different PSC organoids and a control are shown. (BLMC: below lowest measurable concentration; CCA: cholangiocarcinoma; PSC: primary sclerosing cholangitis; ns: not significant) The means with standard deviations are displayed. (*) indicates p value < 0.05.

3.5. STAT3 Is Highly Expressed in the Tumor and Immune Microenvironment of Cholangiocarcinoma

To determine STAT3 expression in cholangiocarcinoma, we downloaded and analyzed data from the TGCA cohort bile duct cancer CHOL (n = 45; 36 primary tumor and 9 normal tissue samples) from the platform XENA [42]. STAT3 is highly expressed in both tumor and normal tissues, and STAT3 expression was expressed at a significantly higher concentration in non-neoplastic tissue samples compared to tumor tissue samples (mean log2 expression: 12.66; IQR: 12.4–12.7 vs. 11.91, IQR: 11.68–12; $p = 0.0015$) (Figure 4E). To evaluate the STAT3 expression at the protein level, we analyzed our BTC cohort after IHC staining. In most samples, STAT3 protein expression was detected both in the tumor compartment (56%, n = 31) and in the immune microenvironment (69%, n = 38) (Supplementary Table S9). The lowest rate of positive STAT3 tumor samples was found in the iCCA group and in the GBC subgroup (50%), while the highest rate was found in the dCCA group (71%). The difference was not statistically significant ($p = 0.7$) (Figure 4D). The rate of STAT3 immune cell positivity was higher in GBC, dCCA, and phCCA compared to iCCA, but the difference was not statistically significant (83%, 80%, and 79%, respectively, vs. 50%, $p = 0.2$) (Figure 4D, Supplementary Table S9).

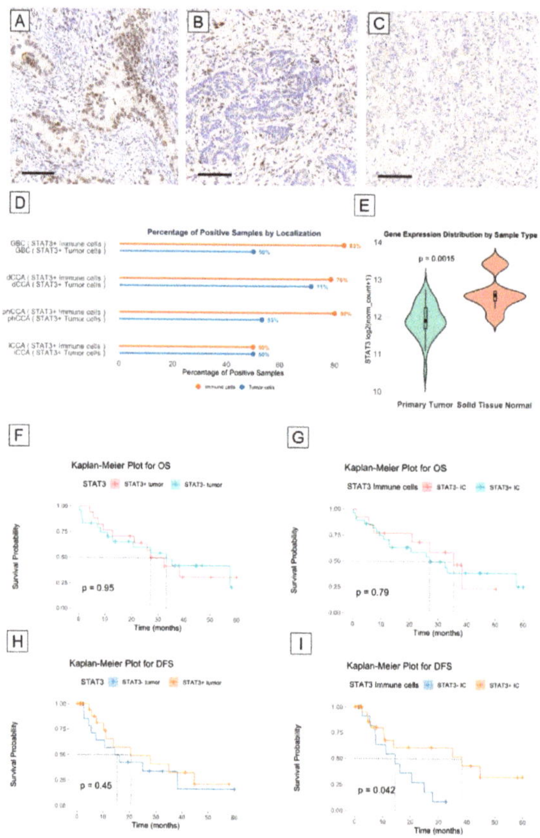

Figure 4. Exemplary images of STAT3+ tumor cells (**A**), STAT3+ immune cells (**B**), and STAT3– tumor cells in the immune microenvironment (**C**). Bar graphs representing the percentage of STAT3+ tumor cells (blue bars) and STAT3+ immune cells (orange bars), grouped by CCA localization (**D**). Violin plot

of gene expression distribution in CCA tumor samples (green) and normal adjacent tissue (orange) from the TGCA-Chol cohort. Median, interquartile range (IQR), and smallest/largest observation greater than or equal to lower/upper hinge—1.5 × IQR (**E**). Kaplan–Meier plot for overall survival according to STAT3 immunoreactivity (low vs. high) in tumor cells (**F**) and in immune cells (**G**). Kaplan–Meier plot for disease-free survival according to STAT3 immunoreactivity (low vs. high) in tumor cells (**H**) and in immune cells (**I**). (GBC: gallbladder cancer; dCCA: distal cholangiocarcinoma; phCCA: perihilar cholangiocarcinoma; iCCA: intrahepatic cholangiocarcinoma; OS: overall survival; DFS: disease-free survival). Scale bar = 100 μm.

3.6. STAT3 Expression May Correlate with Longer Disease-Free Survival in Cholangiocarcinoma Patients

To explore whether STAT3 expression has a role as a prognostic factor, we first compared the overall and disease-free survival (OS and DFS) curves of patients with STAT3+ and STAT− tumor compartments, as well as with a STAT3+ and STAT− immune cell microenvironment. Patients' characteristics according to STAT3 expression are summarized in Supplementary Tables S10 and S11. While we did not observe a difference in OS and DFS between patients with or without STAT3 expression in the tumor compartment, we did observe a significant difference in DFS between patients with STAT3+ and STAT3− expression in the immune cell microenvironment. In fact, the median DFS was longer for patients with a STAT3+ immune cell microenvironment compared to patients with a STAT3− immune cell microenvironment (38.2 vs. 14.5 months, log-rank test $p = 0.042$) (Figure 4F–I). To analyze the role of confounders, we built a Cox proportional hazards model, which ultimately did not confirm the role of STAT3+ expression in the immune microenvironment as an independent prognostic factor for DFS in our cohort (Supplementary Table S12).

4. Discussion

In this study, we successfully established an organoid model from biopsies obtained from PSC patients through endoscopic retrograde cholangiography. We were able to maintain long-term cultures of cholangiocyte organoids from all samples. In addition, we used this model to test a JAK inhibitor. The ability to maintain long-term cholangiocyte organoid cultures from all patient samples highlights the reproducibility of this approach. Furthermore, we demonstrated the feasibility of using this model for novel therapeutic approaches by evaluating the effects of baricitinib as a proof of concept. Organoids provide a powerful tool for disease modeling and personalized medicine because they retain key characteristics of the original tissue, including cellular heterogeneity and functionality [34,43,44]. This is particularly relevant for PSC and other biliary diseases, including CCA, where patient-derived organoids can better recapitulate in vivo responses to treatment and could serve as an intermediate step between in vitro drug screening and in vivo validation [34,45–48]. Previous studies have demonstrated the feasibility of generating reliable organoid models from bile samples, bile duct biopsies, and tissue samples from PSC patients [45,49–52]. These models of PSC that rely on bile fluid and brush cytology may not accurately reflect the localized inflammatory and fibrotic processes occurring at the site of dominant strictures. One of the key strengths of our study is the use of cholangioscopy-guided biopsies of the main strictures of PSC patients to establish organoid cultures. This allows for targeted sampling of areas most affected by chronic inflammation and fibrosis. Previous models derived from bile fluid or donor tissue may not fully represent the cellular and molecular landscape of diseased bile ducts. Our minimally invasive approach captures site-specific epithelial changes within PSC main strictures. In our study, we used endoscopic retrograde cholangiography (ERC) followed by cholangioscopy in nine out of ten cases to precisely identify the location of the main stricture. This enhances the translational relevance of the model and provides a robust platform to study PSC biology, early tumorigenesis and response to therapeutics. The rationale for targeting main strictures in PSC patients stems from their increased risk of malignancy. While strictures in PSC are often benign due to chronic inflammation and fibrosis, they are also the primary sites where cholangiocarcinoma (CCA) can develop [53,54]. By performing cholangioscopy-guided biopsies at the identified main stricture, the aim is to enhance the diagnostic relevance and ensure that the organoids reflect the pathological characteristics of high-risk biliary lesions. With this method, we identified high-risk mutations in two out of ten samples. This approach is more invasive than simple bile sampling, which does not require more biopsies than clinically necessary. However, we did not observe any biopsy-related complications, such as perforation and bleeding, in our cohort.

In our study, we measured the cytokine concentration in the supernatant from the organoids of PSC and CCA samples to determine if the method is suitable for testing novel therapies like baricitinib. Baricitinib is a first-generation, non-selective JAK inhibitor that has been approved for chronic inflammatory conditions like rheumatoid arthritis and colitis ulcerosa [55]. PSC is characterized by chronic inflammation, a condition that may also be driven by activation of the JAK/STAT3 pathway, as demonstrated by recent studies [56–58]. Currently, no approved medication exists for PSC that can modify disease progression or improve overall survival. Therefore, in vivo exploration of novel therapeutic targets remains pivotal.

Aberrant activation of STAT3 has been observed in several cancers, including cholangiocarcinoma [59]. Chronic inflammation is a known risk factor for the development of cholangiocarcinoma; STAT3 is involved in signaling pathways associated with inflammation, and its activation may contribute to the inflammatory microenvironment that promotes tumorigenesis [60]. In our cohort, we found expression of STAT3 protein in most tumors (53%). Interestingly, in a larger proportion of tumors (69%), the immune cells of the tumor microenvironment expressed the STAT3 protein. STAT3 positivity in the immune cell compartment was found to be associated with longer disease-free survival in our cohort.

STAT3 has been associated with pro-tumorigenic activities in many cancers. In our cohort, we found that STAT3 in the immune cell compartment was positively associated with better disease-free survival. However, in our cohort, STAT3 was not identified as an independent prognostic factor, suggesting that its impact on DFS may be influenced by additional factors. In particular, the distribution of tumor subtypes may be an important confounding factor in our study. Specifically, the STAT3+ group in our cohort contained a lower proportion of intrahepatic cholangiocarcinoma (iCCA), which generally have a poorer prognosis compared to extrahepatic subtypes. This imbalance may have contributed to the observed association between STAT3 positivity in the immune microenvironment and improved disease-free survival. Therefore, this prognostic finding should be interpreted considering this potential confounding factor. Although previous studies have associated STAT3 expression, particularly in tumor cells, with adverse clinical outcomes, our results associate STAT3 expression in the immune microenvironment with better outcomes. This may reflect a distinct biological role of STAT3 in immune regulation rather than tumor proliferation or invasion, providing a possible explanation for the observed discrepancy with the existing literature. In fact, the IL-6/STAT3 pathway has been shown to have anti-tumorigenic effects in certain conditions [61]. Therefore, it is plausible that STAT3 activation in immune cells (such as T cells, dendritic cells, and macrophages) may enhance rather than suppress anti-tumor immune responses.

Mutation analysis in liver tumor organoids revealed mutations in key cancer-associated genes, such as TP53 and KRAS, supporting their relevance for studying key molecular drivers of biliary tract cancers and drug response modeling. Interestingly, two

PSC-derived organoids also harbored mutations in KRAS and TP53, which are typically associated with malignant transformation. The presence of these mutations in non-malignant PSC tissue may indicate a higher risk of progression to cholangiocarcinoma in these patients. This finding underscores the potential utility of organoid-based genomic screening to identify high-risk individuals and guide personalized surveillance strategies.

To assess whether the effect of a JAK inhibition can be tested in our CCA organoid model, we treated an organoid with baricitinib. Interestingly, we found a significant reduction in IL-6 and MCP-1 secretion after 48 h of treatment. IL-6 is a stress-response-related cytokine that has been found to be elevated in serum samples of CCA patients [62]. Previous studies have suggested that inhibition of the IL-6 signaling pathway may represent a potential therapeutic strategy in CCA [63,64]. MCP-1 (also known as CCL2, chemokine (C-C motif) ligand 2) overexpression has been linked to tumor growth, angiogenesis, and poor prognosis across multiple cancer types, and elevated serum MCP-1 correlates with tumor stages in various malignancies [65]. In addition, there is evidence that STAT3 activation can lead to MCP-1 overexpression in some tumors, creating a positive feedback loop, thereby facilitating cancer progression [66]. A recent study found that neutralization of CC2 can improve survival and remodel the tumor microenvironment in isocitrate dehydrogenase 1 (IDH1)-mutant cholangiocarcinoma (CCA), a highly aggressive sub-type of CCA [67]. Thus, JAK inhibition could represent an alternative therapeutic option for targeting the IL-6/JAK/STAT3 and the CCL2 signaling pathways.

IL-6 secretion was undetectable in PSC-derived organoids both before and after baricitinib treatment. This may reflect an inherently low basal level of IL-6 production in PSC cholangiocytes, which is consistent with their non-malignant nature, in contrast to the elevated inflammatory signaling observed in CCA. Alternatively, the lack of detectable IL-6 may be due to technical limitations, with cytokine concentrations falling below the assay detection threshold. Further studies, for example, with inflammatory stimulation or more sensitive detection methods, will be required to clarify this issue.

In addition, we performed exploratory tests with commonly used agents to assess the feasibility of using the organoid model for drug screening. Although the number of organoid lines was limited, we observed a variable but consistent decrease in viability following treatment in different liver tumor-derived organoids. These preliminary findings support the potential of patient-derived organoids as a platform for personalized therapeutic testing, although further validation in larger cohorts and in vivo models is required.

Despite these promising findings, several limitations should be considered. First, our PSC organoid model does not recapitulate immune components of the microenvironment. Future studies should explore the integration of co-cultured organoid systems with fibroblasts or immune cells to improve physiological relevance. In addition, we did not assess STAT3 expression in our patient-derived organoid models, particularly in relation to chemosensitivity and cytokine secretion. Future studies will evaluate whether STAT3-positive organoids exhibit differential responses to therapeutic agents or altered cytokine profiles, potentially providing a predictive biomarker for treatment response. Finally, while we tested a single JAK inhibitor, future investigations should be expanded to evaluate combination therapies or other targeted agents.

Supplementary Materials: The following supporting information can be downloaded at: https://www.mdpi.com/article/10.3390/biomedicines13051083/s1, Table S1: Wash medium composition; Table S2: Digestive medium composition; Table S3: Basic organoid medium composition; Table S4: Organoid isolation media composition; Table S5: Organoid expansion media composition; Table S6: Organoid tumoroid medium composition; Table S7: ERC-derived organoid immunoreactivity for Cytokeratin-7 after IHC staining (CK7: Cytokeratin-7); Table S8: Mutations found in the PSC organoid cohort and CCA organoids cohort; Table S9: STAT3 expression in tumor and immune cell microenvironment of the cholangiocarcinoma cohort; Table S10: BTC patients´ cohort divided in subgroup according to tumor cell STAT3 expression (positive vs negative) 1: Median (25% percentile, 75% percentile); n (%); 2: Wilcoxon rank sum test; Pearson's Chi-squared test; Fisher's exact test; Table S11: BTC patient cohort divided in subgroup according to tumor cell STAT3 expression (positive vs. negative) 1: Median (25% percentile, 75% percentile); n (%); 2: Wilcoxon rank sum test; Pearson's Chi-squared test; Fisher's exact test; Table S12: Multivariate analysis for disease-free survival in the CCA cohort.

Author Contributions: Conceptualization, V.B. and H.M.; methodology, M.I.T., V.B., and M.H.; formal analysis, C.B., L.K.E., and V.B.; investigation, C.B., V.B., B.L., and L.K.E.; resources, L.D., B.L., M.I.T., G.K., M.A.G.-C., T.J.W., and D.J.K.; data curation, V.B., T.Z., and C.B.; writing—original draft preparation, C.B.; writing—review and editing, V.B., M.I.T., L.D., D.J.K., T.J.W., B.L., and H.M.; visualization, V.B. and C.B.; supervision, V.B., M.I.T., H.M., and J.C.K. All authors have read and agreed to the published version of the manuscript.

Funding: This study was funded by a grant to V.B. from the Else Kröner-Fresenius Stiftung (grant number: 2014_Kolleg.05).

Institutional Review Board Statement: The study was conducted in accordance with the Declaration of Helsinki and approved by the ethics committee of the Bonn University Hospital (protocol no. 417/17, 27/18, and 233/20).

Informed Consent Statement: Informed consent was obtained from all subjects involved in the study.

Data Availability Statement: The datasets used and analyzed during the current study are available from the corresponding author on reasonable request.

Acknowledgments: The authors would like to thank Kerstin Fuchs, Carsten Golletz, Seher Aktekin, and Susanne Steiner for their help and technical support.

Conflicts of Interest: The authors declare no conflicts of interest. The funders had no role in the design of the study; in the collection, analyses, or interpretation of data; in the writing of the manuscript; or in the decision to publish the results.

Abbreviations

The following abbreviations are used in this manuscript:

PSC	Primary sclerosing cholangitis
BTC	Biliary tract cancer
CCA	Cholangiocarcinoma
iCCA	Intrahepatic cholangiocarcinoma
phCCA	Perihilar cholangiocarcinoma
dCCA	Distal cholangiocarcinoma
GBC	Gallbladder cancer
ERC	Endoscopic retrograde cholangiography
EMT	Epithelial-to-mesenchymal transition
FFPE	Formalin-fixed paraffin-embedded
ECO	extrahepatic cholangiocyte organoid
JAK	Janus kinase
IL-x	Interleukin
MCP-1	Monocyte chemoattractant protein 1
TNM	Tumor, node, metastasis
TMA	Tissue microarray
H&E	Hematoxylin–eosin
ECO	Extrahepatic cholangiocyte organoid
HCC	Hepatocellular carcinoma

CRCm	Colorectal metastasis
TGCA	The Cancer Genome Atlas
IDH1	Isocitrate dehydrogenase 1
STAT3	Signal transducer and activator of transcription 3
IBD	Inflammatory bowel disease
OS	Overall survival
DFS	Disease-free survival

References

1. Lazaridis, K.N.; LaRusso, N.F. Primary Sclerosing Cholangitis. *N. Engl. J. Med.* **2016**, *375*, 1161–1170. [CrossRef]
2. Dean, G.; Hanauer, S.; Levitsky, J. The Role of the Intestine in the Pathogenesis of Primary Sclerosing Cholangitis: Evidence and Therapeutic Implications. *Hepatology* **2020**, *72*, 1127–1138. [CrossRef]
3. Trivedi, P.J.; Bowlus, C.L.; Yimam, K.K.; Razavi, H.; Estes, C. Epidemiology, Natural History, and Outcomes of Primary Sclerosing Cholangitis: A Systematic Review of Population-Based Studies. *Clin. Gastroenterol. Hepatol.* **2022**, *20*, 1687–1700.e4. [CrossRef] [PubMed]
4. Molodecky, N.A.; Kareemi, H.; Parab, R.; Barkema, H.W.; Quan, H.; Myers, R.P.; Kaplan, G.G. Incidence of Primary Sclerosing Cholangitis: A Systematic Review and Meta-Analysis. *Hepatology* **2011**, *53*, 1590–1599. [CrossRef]
5. Takakura, W.R.; Tabibian, J.H.; Bowlus, C.L. The Evolution of Natural History of Primary Sclerosing Cholangitis. *Curr. Opin. Gastroenterol.* **2017**, *33*, 71–77. [CrossRef] [PubMed]
6. Lunder, A.K.; Hov, J.R.; Borthne, A.; Gleditsch, J.; Johannesen, G.; Tveit, K.; Viktil, E.; Henriksen, M.; Hovde, Ø.; Huppertz-Hauss, G.; et al. Prevalence of Sclerosing Cholangitis Detected by Magnetic Resonance Cholangiography in Patients with Long-Term Inflammatory Bowel Disease. *Gastroenterology* **2016**, *151*, 660–669.e4. [CrossRef]
7. Folseraas, T.; Boberg, K.M. Cancer Risk and Surveillance in Primary Sclerosing Cholangitis. *Clin. Liver Dis.* **2016**, *20*, 79–98. [CrossRef] [PubMed]
8. Bergquist, A.; Ekbom, A.; Olsson, R.; Kornfeldt, D.; Lööf, L.; Danielsson, Å.; Hultcrantz, R.; Lindgren, S.; Prytz, H.; Sandberg-Gertzén, H.; et al. Hepatic and Extrahepatic Malignancies in Primary Sclerosing Cholangitis. *J. Hepatol.* **2002**, *36*, 321–327. [CrossRef]
9. Claessen, M.M.H.; Vleggaar, F.P.; Tytgat, K.M.A.J.; Siersema, P.D.; van Buuren, H.R. High Lifetime Risk of Cancer in Primary Sclerosing Cholangitis. *J. Hepatol.* **2009**, *50*, 158–164. [CrossRef]
10. Hrad, V.; Abebe, Y.; Ali, S.H.; Velgersdyk, J.; Al Hallak, M.; Imam, M. Risk and Surveillance of Cancers in Primary Biliary Tract Disease. *Gastroenterol Res. Pract.* **2016**, *2016*, 3432640. [CrossRef]
11. Boonstra, K.; Weersma, R.K.; van Erpecum, K.J.; Rauws, E.A.; Spanier, B.W.M.; Poen, A.C.; van Nieuwkerk, K.M.; Drenth, J.P.; Witteman, B.J.; Tuynman, H.A.; et al. Population-Based Epidemiology, Malignancy Risk, and Outcome of Primary Sclerosing Cholangitis. *Hepatology* **2013**, *58*, 2045–2055. [CrossRef]
12. Albert, J.; Arvanitakis, M.; Chazouilleres, O.; Dumonceau, J.-M.; Färkkilä, M.; Fickert, P.; Hirschfield, G.M.; Laghi, A.; Marzioni, M.; Fernandez, M.; et al. Role of Endoscopy in Primary Sclerosing Cholangitis: European Society of Gastrointestinal Endoscopy (ESGE) and European Association for the Study of the Liver (EASL) Clinical Guideline. *J. Hepatol.* **2017**, *66*, 1265–1281. [CrossRef]
13. Valle, J.W.; Kelley, R.K.; Nervi, B.; Oh, D.Y.; Zhu, A.X. Biliary Tract Cancer. *Lancet* **2021**, *397*, 428–444. [CrossRef]
14. Forner, A.; Vidili, G.; Rengo, M.; Bujanda, L.; Ponz-Sarvisé, M.; Lamarca, A. Clinical Presentation, Diagnosis and Staging of Cholangiocarcinoma. *Liver Int.* **2019**, *39*, 98–107. [CrossRef] [PubMed]
15. Vogel, A.; Bridgewater, J.; Edeline, J.; Kelley, R.K.; Klümpen, H.J.; Malka, D.; Primrose, J.N.; Rimassa, L.; Stenzinger, A.; Valle, J.W.; et al. Biliary Tract Cancer: ESMO Clinical Practice Guideline for Diagnosis, Treatment and Follow-Up. *Ann. Oncol.* **2023**, *34*, 127–140. [CrossRef] [PubMed]
16. Brindley, P.J.; Bachini, M.; Ilyas, S.I.; Khan, S.A.; Loukas, A.; Sirica, A.E.; Teh, B.T.; Wongkham, S.; Gores, G.J. Cholangiocarcinoma. *Nat. Rev. Dis. Primers* **2021**, *7*, 65. [CrossRef]
17. Khan, S.A.; Tavolari, S.; Brandi, G. Cholangiocarcinoma: Epidemiology and Risk Factors. *Liver Int.* **2019**, *39*, 19–31. [CrossRef]
18. Sithithaworn, P.; Yongvanit, P.; Duenngai, K.; Kiatsopit, N.; Pairojkul, C. Roles of Liver Fluke Infection as Risk Factor for Cholangiocarcinoma. *J. Hepatobiliary Pancreat. Sci.* **2014**, *21*, 301–308. [CrossRef]
19. Banales, J.M.; Cardinale, V.; Carpino, G.; Marzioni, M.; Andersen, J.B.; Invernizzi, P.; Lind, G.E.; Folseraas, T.; Forbes, S.J.; Fouassier, L.; et al. Expert Consensus Document: Cholangiocarcinoma: Current Knowledge and Future Perspectives Consensus Statement from the European Network for the Study of Cholangiocarcinoma (ENS-CCA). *Nat. Rev. Gastroenterol. Hepatol.* **2016**, *13*, 261–280. [CrossRef]
20. Clements, O.; Eliahoo, J.; Kim, J.U.; Taylor-Robinson, S.D.; Khan, S.A. Risk Factors for Intrahepatic and Extrahepatic Cholangiocarcinoma: A Systematic Review and Meta-Analysis. *J. Hepatol.* **2020**, *72*, 95–103. [CrossRef]

21. Bray, F.; Laversanne, M.; Sung, H.; Ferlay, J.; Siegel, R.L.; Soerjomataram, I.; Jemal, A. Global Cancer Statistics 2022: GLOBOCAN Estimates of Incidence and Mortality Worldwide for 36 Cancers in 185 Countries. *CA Cancer J. Clin.* **2024**, *74*, 229–263. [CrossRef] [PubMed]
22. Piovani, D.; Nikolopoulos, G.K.; Aghemo, A.; Lleo, A.; Alqahtani, S.A.; Hassan, C.; Repici, A.; Bonovas, S. Environmental Risk Factors for Gallbladder Cancer: Field-Wide Systematic Review and Meta-Analysis. *Clin. Gastroenterol. Hepatol.* **2024**. [CrossRef]
23. Hu, Y.; Dong, Z.; Liu, K. Unraveling the Complexity of STAT3 in Cancer: Molecular Understanding and Drug Discovery. *J. Exp. Clin. Cancer Res.* **2024**, *43*, 23. [CrossRef] [PubMed]
24. Liang, B.; Li, S.Y.; Gong, H.Z.; Wang, L.X.; Lu, J.; Zhao, Y.X.; Gu, N. Clinicopathological and Prognostic Roles of STAT3 and Its Phosphorylation in Glioma. *Dis. Markers* **2020**, *2020*, 8833885. [CrossRef]
25. Denley, S.M.; Jamieson, N.B.; McCall, P.; Oien, K.A.; Morton, J.P.; Carter, C.R.; Edwards, J.; McKay, C.J. Activation of the IL-6R/Jak/Stat Pathway Is Associated with a Poor Outcome in Resected Pancreatic Ductal Adenocarcinoma. *J. Gastrointest. Surg.* **2013**, *17*, 887–898. [CrossRef] [PubMed]
26. Tong, M.; Wang, J.; Jiang, N.; Pan, H.; Li, D. Correlation between P-STAT3 Overexpression and Prognosis in Lung Cancer: A Systematic Review and Meta-Analysis. *PLoS ONE* **2017**, *12*, e0182282. [CrossRef]
27. Sayed, D.; Badrawy, H.; Gaber, N.; Khalaf, M.R. P-Stat3 and Bcr/Abl Gene Expression in Chronic Myeloid Leukemia and Their Relation to Imatinib Therapy. *Leuk. Res.* **2014**, *38*, 243–250. [CrossRef]
28. Chang, Q.; Bournazou, E.; Sansone, P.; Berishaj, M.; Gao, S.P.; Daly, L.; Wels, J.; Theilen, T.; Granitto, S.; Zhang, X.; et al. The IL-6/JAK/Stat3 Feed-Forward Loop Drives Tumorigenesis and Metastasis. *Neoplasia* **2013**, *15*, 848–862. [CrossRef]
29. Zhang, H.X.; Yang, P.L.; Li, E.M.; Xu, L.Y. STAT3beta, a Distinct Isoform from STAT3. *Int. J. Biochem. Cell Biol.* **2019**, *110*, 130–139. [CrossRef]
30. Huynh, J.; Chand, A.; Gough, D.; Ernst, M. Therapeutically Exploiting STAT3 Activity in Cancer—Using Tissue Repair as a Road Map. *Nat. Rev. Cancer* **2019**, *19*, 82–96. [CrossRef]
31. Luo, Z.; Jegga, A.G.; Bezerra, J.A. Gene-Disease Associations Identify a Connectome with Shared Molecular Pathways in Human Cholangiopathies. *Hepatology* **2018**, *67*, 676–689. [CrossRef] [PubMed]
32. Xin-Wei, Y.; Liang, L.; Guo-Jun, H.; Xin-Zhou, Y.; Qin-Guo, X.; Lei, C.; Bao-Hua, Z.; Feng, S. STAT3 Overexpression Promotes Metastasis in Intrahepatic Cholangiocarcinoma and Correlates Negatively with Surgical Outcome. *Oncotarget* **2016**, *8*, 7710. [CrossRef]
33. Zhang, C.; Xu, H.; Zhou, Z.; Tian, Y.; Cao, X.; Cheng, G.; Liu, Q. Blocking of the EGFR-STAT3 Signaling Pathway through Afatinib Treatment Inhibited the Intrahepatic Cholangiocarcinoma. *Exp. Ther. Med.* **2018**, *15*, 4995–5000. [CrossRef]
34. Broutier, L.; Mastrogiovanni, G.; Verstegen, M.M.A.; Francies, H.E.; Gavarró, L.M.; Bradshaw, C.R.; Allen, G.E.; Arnes-Benito, R.; Sidorova, O.; Gaspersz, M.P.; et al. Human Primary Liver Cancer–Derived Organoid Cultures for Disease Modeling and Drug Screening. *Nat. Med.* **2017**, *23*, 1424–1435. [CrossRef]
35. Kassambara, A.; Kosinski, M.; Biecek, P. Survminer: Drawing Survival Curves Using "Ggplot2". 2020. Available online: https://cran.r-project.org/web/packages/survminer/survminer.pdf (accessed on 15 January 2025).
36. R Core Team. *R: A Language and Environment for Statistical Computing*; R Foundation for Statistical Computing: Vienna, Austria, 2022.
37. Therneau, T. A Package for Survival Analysis in R. R Package Version 3.5-5. 2023. Available online: https://cran.r-project.org/web/packages/survival/vignettes/survival.pdf (accessed on 15 January 2025).
38. Therneau, T.M.; Grambsch, P.M. *Modeling Survival Data: Extending the Cox Model*; Springer: New York, NY, USA, 2000; ISBN 0-387-98784-3.
39. Sjoberg, D.D.; Whiting, K.; Curry, M.; Lavery, J.A.; Larmarange, J. Reproducible Summary Tables with the Gtsummary Package. *R J.* **2021**, *13*, 570–580. [CrossRef]
40. Iannone, R.; Cheng, J.; Schloerke, B.; Hughes, E.; Lauer, A.; Seo, J.; Brevoort, K.; Roy, O. Gt: Easily Create Presentation-Ready Display Tables. 2024. Available online: https://gt.rstudio.com/ (accessed on 15 January 2025).
41. Wickham, H. Ggplot2: Elegant Graphics for Data Analysis. *J. Stat. Softw.* **2009**, *35*, 160–167.
42. Goldman, M.J.; Craft, B.; Hastie, M.; Repečka, K.; McDade, F.; Kamath, A.; Banerjee, A.; Luo, Y.; Rogers, D.; Brooks, A.N.; et al. Visualizing and Interpreting Cancer Genomics Data via the Xena Platform. *Nat. Biotechnol.* **2020**, *38*, 675–678. [CrossRef]
43. Fatehullah, A.; Tan, S.H.; Barker, N. Organoids as an in Vitro Model of Human Development and Disease. *Nat. Cell Biol.* **2016**, *18*, 246–254. [CrossRef] [PubMed]
44. Sachs, N.; de Ligt, J.; Kopper, O.; Gogola, E.; Bounova, G.; Weeber, F.; Balgobind, A.V.; Wind, K.; Gracanin, A.; Begthel, H.; et al. A Living Biobank of Breast Cancer Organoids Captures Disease Heterogeneity. *Cell* **2018**, *172*, 373–386.e10. [CrossRef]
45. Soroka, C.J.; Assis, D.N.; Alrabadi, L.S.; Roberts, S.; Cusack, L.; Jaffe, A.B.; Boyer, J.L. Bile-Derived Organoids From Patients With Primary Sclerosing Cholangitis Recapitulate Their Inflammatory Immune Profile. *Hepatology* **2019**, *70*, 871–882. [CrossRef]

46. Nuciforo, S.; Fofana, I.; Matter, M.S.; Blumer, T.; Calabrese, D.; Boldanova, T.; Piscuoglio, S.; Wieland, S.; Ringnalda, F.; Schwank, G.; et al. Organoid Models of Human Liver Cancers Derived from Tumor Needle Biopsies. *Cell Rep.* **2018**, *24*, 1363–1376. [CrossRef] [PubMed]
47. Saito, Y.; Muramatsu, T.; Kanai, Y.; Ojima, H.; Sukeda, A.; Hiraoka, N.; Arai, E.; Sugiyama, Y.; Matsuzaki, J.; Uchida, R.; et al. Establishment of Patient-Derived Organoids and Drug Screening for Biliary Tract Carcinoma. *Cell Rep.* **2019**, *27*, 1265–1276.e4. [CrossRef] [PubMed]
48. Van De Wetering, M.; Francies, H.E.; Francis, J.M.; Bounova, G.; Iorio, F.; Pronk, A.; Van Houdt, W.; Van Gorp, J.; Taylor-Weiner, A.; Kester, L.; et al. Prospective Derivation of a Living Organoid Biobank of Colorectal Cancer Patients. *Cell* **2015**, *161*, 933–945. [CrossRef] [PubMed]
49. Tabibian, J.H.; Trussoni, C.E.; O'Hara, S.P.; Splinter, P.L.; Heimbach, J.K.; LaRusso, N.F. Characterization of Cultured Cholangiocytes Isolated from Livers of Patients with Primary Sclerosing Cholangitis. *Lab. Investig.* **2014**, *94*, 1126–1133. [CrossRef]
50. Verstegen, M.M.A.; Roos, F.J.M.; Burka, K.; Gehart, H.; Jager, M.; de Wolf, M.; Bijvelds, M.J.C.; de Jonge, H.R.; Ardisasmita, A.I.; van Huizen, N.A.; et al. Human Extrahepatic and Intrahepatic Cholangiocyte Organoids Show Region-Specific Differentiation Potential and Model Cystic Fibrosis-Related Bile Duct Disease. *Sci. Rep.* **2020**, *10*, 21900. [CrossRef]
51. Roos, F.J.M.; Verstegen, M.M.A.; Muñoz Albarinos, L.; Roest, H.P.; Poley, J.W.; Tetteroo, G.W.M.; IJzermans, J.N.M.; van der Laan, L.J.W. Human Bile Contains Cholangiocyte Organoid-Initiating Cells Which Expand as Functional Cholangiocytes in Non-Canonical Wnt Stimulating Conditions. *Front. Cell Dev. Biol.* **2021**, *8*, 630492. [CrossRef]
52. Moreno, A.S.G.; Guicciardi, M.E.; Wixom, A.Q.; Jessen, E.; Yang, J.; Ilyas, S.I.; Bianchi, J.K.; Pinto E Vairo, F.; Lazaridis, K.N.; Gores, G.J. IL-17 Signaling in Primary Sclerosing Cholangitis Patient-Derived Organoids. *Hepatol. Commun.* **2024**, *8*, e0454. [CrossRef]
53. Chapman, R.W.; Williamson, K.D. Are Dominant Strictures in Primary Sclerosing Cholangitis a Risk Factor for Cholangiocarcinoma? *Curr. Hepatol Rep.* **2017**, *16*, 124–129. [CrossRef]
54. Chapman, M.H.; Webster, G.J.M.; Bannoo, S.; Johnson, G.J.; Wittmann, J.; Pereira, S.P. Cholangiocarcinoma and Dominant Strictures in Patients with Primary Sclerosing Cholangitis; a 25 Year Single Centre Experience. *Eur. J. Gastroenterol. Hepatol.* **2012**, *24*, 1051. [CrossRef]
55. Chikhoune, L.; Poggi, C.; Moreau, J.; Dubucquoi, S.; Hachulla, E.; Collet, A.; Launay, D. JAK Inhibitors (JAKi): Mechanisms of Action and Perspectives in Systemic and Autoimmune Diseases. *Rev. Med. Interne* **2025**, *46*, 89–106. [CrossRef]
56. Asuri, S.; McIntosh, S.; Taylor, V.; Rokeby, A.; Kelly, J.; Shumansky, K.; Field, L.L.; Yoshida, E.M.; Arbour, L. Primary Biliary Cholangitis in British Columbia First Nations: Clinical Features and Discovery of Novel Genetic Susceptibility Loci. *Liver Int.* **2018**, *38*, 940–948. [CrossRef] [PubMed]
57. Dold, L.; Frank, L.; Lutz, P.; Kaczmarek, D.J.; Krämer, B.; Nattermann, J.; Weismüller, T.J.; Branchi, V.; Toma, M.; Gonzalez-Carmon, M.; et al. IL-6–Dependent STAT3 Activation and Induction of Proinflammatory Cytokines in Primary Sclerosing Cholangitis. *Clin. Transl. Gastroenterol.* **2023**, *14*, e00603. [CrossRef] [PubMed]
58. Poch, T.; Krause, J.; Casar, C.; Liwinski, T.; Glau, L.; Kaufmann, M.; Ahrenstorf, A.E.; Hess, L.U.; Ziegler, A.E.; Martrus, G.; et al. Single-Cell Atlas of Hepatic T Cells Reveals Expansion of Liver-Resident Naive-like CD4+ T Cells in Primary Sclerosing Cholangitis. *J. Hepatol.* **2021**, *75*, 414–423. [CrossRef]
59. Dokduang, H.; Techasen, A.; Namwat, N.; Khuntikeo, N.; Pairojkul, C.; Murakami, Y.; Loilome, W.; Yongvanit, P. STATs Profiling Reveals Predominantly-Activated STAT3 in Cholangiocarcinoma Genesis and Progression. *J. Hepatobiliary Pancreat. Sci.* **2014**, *21*, 767–776. [CrossRef]
60. Cadamuro, M.; Strazzabosco, M. Inflammatory Pathways and Cholangiocarcinoma Risk Mechanisms and Prevention. *Adv. Cancer Res.* **2022**, *156*, 39–73. [CrossRef] [PubMed]
61. Shriki, A.; Lanton, T.; Sonnenblick, A.; Levkovitch-Siany, O.; Eidelshtein, D.; Abramovitch, R.; Rosenberg, N.; Pappo, O.; Elgavish, S.; Nevo, Y.; et al. Multiple Roles of Il6 in Hepatic Injury, Steatosis, and Senescence Aggregate to Suppress Tumorigenesis. *Cancer Res.* **2021**, *81*, 4766–4777. [CrossRef]
62. Cheon, Y.K.; Cho, Y.D.; Moon, J.H.; Jang, J.Y.; Kim, Y.S.; Kim, Y.S.; Lee, M.S.; Lee, J.S.; Shim, C.S. Diagnostic Utility of Interleukin-6 (IL-6) for Primary Bile Duct Cancer and Changes in Serum IL-6 Levels Following Photodynamic Therapy. *Am. J. Gastroenterol.* **2007**, *102*, 2164–2170. [CrossRef]
63. Kleinegger, F.; Hofer, E.; Wodlej, C.; Golob-Schwarzl, N.; Birkl-Toeglhofer, A.M.; Stallinger, A.; Petzold, J.; Orlova, A.; Krassnig, S.; Reihs, R.; et al. Pharmacologic IL-6Rα Inhibition in Cholangiocarcinoma Promotes Cancer Cell Growth and Survival. *Biochim. Biophys. Acta Mol. Basis Dis.* **2019**, *1865*, 308–321. [CrossRef]
64. Nguyen, M.L.T.; Bui, K.C.; Scholta, T.; Xing, J.; Bhuria, V.; Sipos, B.; Wilkens, L.; Nguyen Linh, T.; Velavan, T.P.; Bozko, P.; et al. Targeting Interleukin 6 Signaling by Monoclonal Antibody Siltuximab on Cholangiocarcinoma. *J. Gastroenterol. Hepatol.* **2021**, *36*, 1334–1345. [CrossRef]
65. Hao, Q.; Vadgama, J.V.; Wang, P. CCL2/CCR2 Signaling in Cancer Pathogenesis. *Cell Commun. Signal.* **2020**, *18*, 82. [CrossRef]

66. Chen, W.; Gao, Q.; Han, S.; Pan, F.; Fan, W. The CCL2/CCR2 Axis Enhances IL-6-Induced Epithelial-Mesenchymal Transition by Cooperatively Activating STAT3-Twist Signaling. *Tumor Biol.* **2015**, *36*, 973–981. [CrossRef] [PubMed]
67. Zabransky, D.J.; Kartalia, E.; Lee, J.W.; Leatherman, J.M.; Charmsaz, S.; Young, S.E.; Chhabra, Y.; Franch-Expósito, S.; Kang, M.; Maru, S.; et al. Tumor-Derived CCL2 Drives Tumor Growth and Immunosuppression in IDH1-Mutant Cholangiocarcinoma. *Hepatology* **2024**. [CrossRef] [PubMed]

biomedicines

MDPI

Article

Correlative Analysis of Tumor-Informed Circulating Tumor DNA (ctDNA) and the Survival Outcomes of Patients with Pancreatic Adenocarcinoma

Yuqi Zhang [1], Abdullah Esmail [1], Hala Hassanain [1], Vikramjit Dhillon [1], Waseem Abdelrahim [2], Ebtesam Al-Najjar [1], Bayan Khasawneh [1] and Maen Abdelrahim [1,*]

1 Section of GI Oncology, Houston Methodist Neal Cancer Center, Houston Methodist Hospital, Houston, TX 77094, USA

2 Michael E. DeBakey HS for Health Professions, Houston, TX 77030, USA

* Correspondence: mabdelrahim@houstonmethodist.org

Abstract: Background: Pancreatic ductal adenocarcinoma (PDAC) is a highly aggressive cancer with poor prognosis due to late-stage diagnosis, limited surgical resectability, and frequent recurrence. Traditional biomarkers like CA19-9 and imaging techniques often fail to detect minimal residual disease (MRD) or early recurrence. Circulating tumor DNA (ctDNA) is a promising non-invasive biomarker that may provide early detection of disease recurrence, offering a potential improvement in patient management. This study aimed to assess the utility of ctDNA as a prognostic tool for PDAC patients, specifically in predicting recurrence and overall survival (OS). **Methods:** This retrospective study analyzed data from 39 PDAC patients who underwent surgery and were monitored for ctDNA levels using Signatera™, a tumor-informed multiplex PCR next-generation sequencing assay. Blood samples were collected both preoperatively and postoperatively, and ctDNA levels were measured to detect MRD. The sensitivity, specificity, positive predictive value (PPV), and negative predictive value (NPV) of ctDNA were compared with CA19-9 in detecting disease recurrence. Clinical outcomes, including progression-free survival (PFS) and OS, were evaluated in relation to ctDNA status. **Results:** Among 39 patients, 153 plasma samples were analyzed, with 17 patients testing positive for ctDNA. Sensitivity of ctDNA in detecting relapse was 91%, compared to 83% for CA19-9, with combined testing reaching 98% sensitivity. ctDNA positivity was associated with significantly shorter OS and PFS, with patients testing negative for ctDNA having a median OS of 37.6 months versus 13.4 months in ctDNA-positive patients ($p = 0.003$). The median time from ctDNA positivity to imaging-confirmed relapse was 81 days. Positive ctDNA was also linked to higher rates of lymphovascular invasion and positive surgical margins, highlighting the aggressive nature of the disease in these patients. **Conclusions:** CtDNA is a highly sensitive and specific biomarker for detecting MRD and predicting recurrence in PDAC patients, offering superior performance over CA19-9. Positive ctDNA results were associated with worse prognosis, including shorter OS and PFS, and may help guide treatment decisions. These findings suggest that ctDNA could be a valuable tool for personalized management in PDAC, though further prospective studies are needed to validate its clinical role in treatment stratification.

Keywords: pancreatic ductal adenocarcinoma; PDAC; circulating tumor DNA; minimal residual disease; MRD; recurrence; progression-free survival; overall survival; tumor-informed assay; biomarker; prognostic marker; CA19-9; adjuvant chemotherapy; non-invasive monitoring; liquid biopsy; personalized treatment; surveillance

1. Introduction

In 2019, there were an estimated 56,770 individuals in the United States who received a new diagnosis of pancreatic cancer, and roughly 46,000 died from the disease [1]. While the mortality rates of stomach and colorectal cancers have been decreasing in the last twenty years, there has been no reduction in mortality rates for pancreatic cancer [1–3]. Currently, pancreatic cancer accounts for approximately 3% of all cancer diagnoses and 7% of cancer-related fatalities [4]. It is expected that by 2030, pancreatic cancer will become the second leading cause of cancer-related mortality [5].

Approximately 80–90% of pancreatic malignancies are categorized as pancreatic ductal adenocarcinomas (PDACs) [2,3,6–8]. A primary factor contributing to the high death rate of PDACs is the prevalence of advanced-stage disease upon diagnosis in most patients. At the time of diagnosis, only a small percentage, specifically 15–20% of patients, are deemed suitable for surgical intervention [9]. In addition, the prognosis for patients who undergo surgery with negative margins is still unfavorable, with a 5-year survival rate ranging from only 10–25% and a median survival of 10–20 months [10]. The National Comprehensive Cancer Network recommends 6-month adjuvant chemotherapy for curative intent after surgery for patients with resectable PDAC [11,12]. Yet, more than 75% of patients experience recurrence following surgery [13]. The current 5-year survival rate for PDAC is 12% in general but drops to 3% for individuals with metastatic disease [4,14]. However, the National Comprehensive Cancer Network (NCCN) panel and other specialized pancreatic cancer centers currently recommend that patients who are very likely to have early signs of metastasis receive neoadjuvant therapy six months prior to any surgical intervention [15]. The American Society of Clinical Oncology (ASCO) guidelines also recommend neoadjuvant therapy for patients with resectable PDAC who cannot undergo immediate surgery [16,17].

Optimization of treatment depends on identification of high-risk patients who are more likely to recur or progress and would benefit most from neoadjuvant and adjuvant systemic therapy. Recent prospective trials have shown that adjuvant treatments are suitable for all patients with resectable or borderline resectable PDAC [15,18,19]. However, the implementation of adjuvant treatments may not be feasible for all patients undergoing surgical resection, primarily due to the high incidence of postoperative complications. Moreover, it is frequently essential to reduce the dose of chemotherapy in patients who have had surgery, potentially resulting in reduced efficacy of the treatment. Neoadjuvant treatments may be a better choice from this point of view because they can be given to a larger group of patients who are planning to have curative procedures and because they have a higher rate of therapeutic efficacy [20]. Currently, clinicians monitor diseases using a combination of clinical symptoms, a tumor marker known as cancer antigen 19-9 (CA 19-9), and computed tomography (CT) or magnetic resonance imaging (MRI) [21]. Still, CA-19-9 levels can rise in both cancer and noncancerous conditions, like biliary inflammation or blockage, which makes it a biomarker that is not very specific [22,23]. Yet traditional imaging techniques are often not sensitive enough to detect minimum residual disease (MRD), defined as the presence of residual cancer cells undetectable by routine imaging or lab tests but potentially leading to relapse [24,25].

Circulating tumor DNA (ctDNA) is a blood biomarker capable of detecting disease recurrence prior detection on imaging studies [26–28]. Liquid biopsy is an emerging technique that enables the collection of tumor material that is circulating in the body without the need for invasive procedures, and its efficacy is being investigated in various forms of gastrointestinal malignancies [29]. It is expected to offer both practical and theoretical advantages over current diagnostic standards [30,31]. The use of cfDNA, ctDNA, tumor-derived exosomes, and circulating tumor cells (CTCs) in PDAC has attracted much attention lately [32,33]. Studies in various cancers have shown that ctDNA is a dependable technique for detecting MRD and informing treatment decisions [34–36]. In this study, we tracked ctDNA levels over time using Signatera^{TM}, a special multiplex PCR next-generation sequencing (NGS) assay. The objective was to verify ctDNA as a disease-monitoring tool that can assess effectiveness of treatment during the perioperative phase, and predict the likelihood of recurrence during surveillance. Furthermore, we aim to confirm the utility of ctDNA as a dependable indicator of recurrence that can be used to improve patient risk stratification throughout treatment.

2. Methodology

2.1. Study Cohort and Sample Collection

This study retrospectively analyzed real-world data of individuals with PDAC and ampullary carcinoma. Data from commercial ctDNA testing conducted between December 2019 to November 2023 were used from a selection of 39 patients diagnosed with PDAC. In the period between June 2020 and December 2023, blood samples were obtained from these patients and stored in a biobank. Each sample underwent ctDNA testing utilizing the tumor-informed assay Signatera^{TM}. The exclusion criteria encompassed patients who did not possess Signatera results, lacked complete and validated clinical data or follow-up, had other histologic subtypes such as pancreatic neuroendocrine tumors, or did not provide informed consent. The individuals who were eligible for the investigation were those who had a confirmed diagnosis of PDAC or ampullary carcinoma (N = 1) as well as longitudinal data on ctDNA and disease-free survival (DFS). Plasma samples were collected both before and after the surgical procedure. The samples were examined for ctDNA. In addition to that, ctDNA analysis was carried out throughout the monitoring period. Initial ctDNA was generally obtained 14 to 534 days after surgery, with a median of 75 days followed by every 1 to 3 months thereafter until death or last follow up.

Clinical and pathological data of every patient were collected. Both the biobank patient samples and the commercial patient samples, which were treated to perioperative care, displayed homogeneity in their biological characteristics. All of the patients, with the exception of the biobank cohort, were treated and monitored according to the expert opinion of the attending physician. Prior to the information regarding the clinical data being disclosed, the statistical analysis plan for ctDNA was developed. Anonymization was performed on the data before it was evaluated. Patient consent was obtained prior to sample collection.

2.2. Personalized mPCR-Based NGS Assay for ctDNA Detection

Signatera^{TM}, a multiplex PCR NGS method, was used to detect and measure tissue-specific DNA. Whole exome sequencing was performed on cell-free DNA from blood samples and tumor blocks from the same patients. To account for germline mutations, both tumor tissue and matched normal tissue (germline) from each patient are sequenced to identify patient-specific mutations and then, using this information, a custom, personalized assay is designed. This assay then targets tumor-specific clonal mutations, filtering out germline mutations. Universal libraries were produced using specialized adapters and end repair, A-tailing, and ligation processes. Libraries were then processed through barcoding, consolidation, mPCR amplification, and sequencing with NGS technology. Plasma samples were considered ctDNA-positive if they displayed a minimum of two different alterations. The quantity of ctDNA was calculated by determining the average number of tumor molecules (MTM) present in one milliliter of plasma.

3. Results

3.1. Patient Characteristics

Plasma samples (n = 153) from patients (N = 39) with PDAC (18 females vs. 21 males) were obtained. The median age at diagnosis was 67 years old (Range: 41–83). Baseline patient characteristics between patients with ctDNA that was initially negative (ctDNA (−)) (N = 22) and initially positive (ctDNA (+)) (N = 17) were comparable (see Table 1).

Table 1. Patient characteristics.

	ctDNA (−)	ctDNA (+)
Male sex	12 (55%)	9 (53%)
Median age	65 (48–82)	68 (40–83)
Site of tumor		
Head	17 (77%)	13 (76%)
Body/tail	5 (23%)	4 (24%)
Underwent resection	19 (86%)	11 (65%)
Margins clear on resection	18 (95%)	7 (64%)
Lymphovascular invasion present	6 (32%)	10 (91%)

A larger portion of ctDNA (−) patients had resectable disease (18 of 21) compared to the ctDNA (+) group (11 out of 17). Of those who underwent surgical resection, a higher percentage of ctDNA (−) patients had negative margins (94% compared to 64%) compared to ctDNA (+) patients. Lymphovascular invasion (LVI) was higher in ctDNA (+) patients (91% compared to 28% in ctDNA (−) patients) in those who underwent resection.

3.2. ctDNA as Detection for MRD

Of the 29 patients who underwent surgery for PDAC, 15 (52%) had ctDNA collected within the MRD window (defined as within 3 months after resection), with 9 of the 15 (60%) having positive ctDNA in the MRD window. All of the nine patients had confirmed progression on later imaging (detection rate 100%). Twenty-six of the twenty-nine received adjuvant chemotherapy with either gemcitabine and abraxane or FOLFIRINOX. Two of the ctDNA (−) patients also received neoadjuvant chemotherapy in addition to adjuvant therapy (see Figure 1).

The sensitivities, specificities, and positive and negative predictive values of the ctDNA test compared to CA19-9 are listed in Table 2. Briefly, in this context, positive predictive value is a measure of how accurately a positive ctDNA result indicates residual PDAC.

Table 2. Sensitivity and specificity of ctDNA.

	ctDNA	CA19-9	Combined
Sensitivity	0.913	0.826	0.98
Specificity	0.813	0.8	0.96
Positive predictive value (PPV)	0.875	0.863	0.97
Negative predictive value (NPV)	0.867	0.75	0.98

Sensitivity and negative predictive value (NPV) of ctDNA in detection of residual or progressive disease as confirmed by CT imaging were higher at 91% (CI 72–99%) and 87% (CI 60–98%), compared to that of CA19-9 (83% (CI 61–95%) and 75% (CI 48–93%)), respectively. Meanwhile, specificities were very similar between ctDNA and CA19-9 testing

(81% (CI 54–96%) vs. 80% (CI 52–96%)). Notably, combining the two tests, sensitivity and specificity increased to 98% and 96%, with corresponding increases in PPV and NPV (97% and 98%).

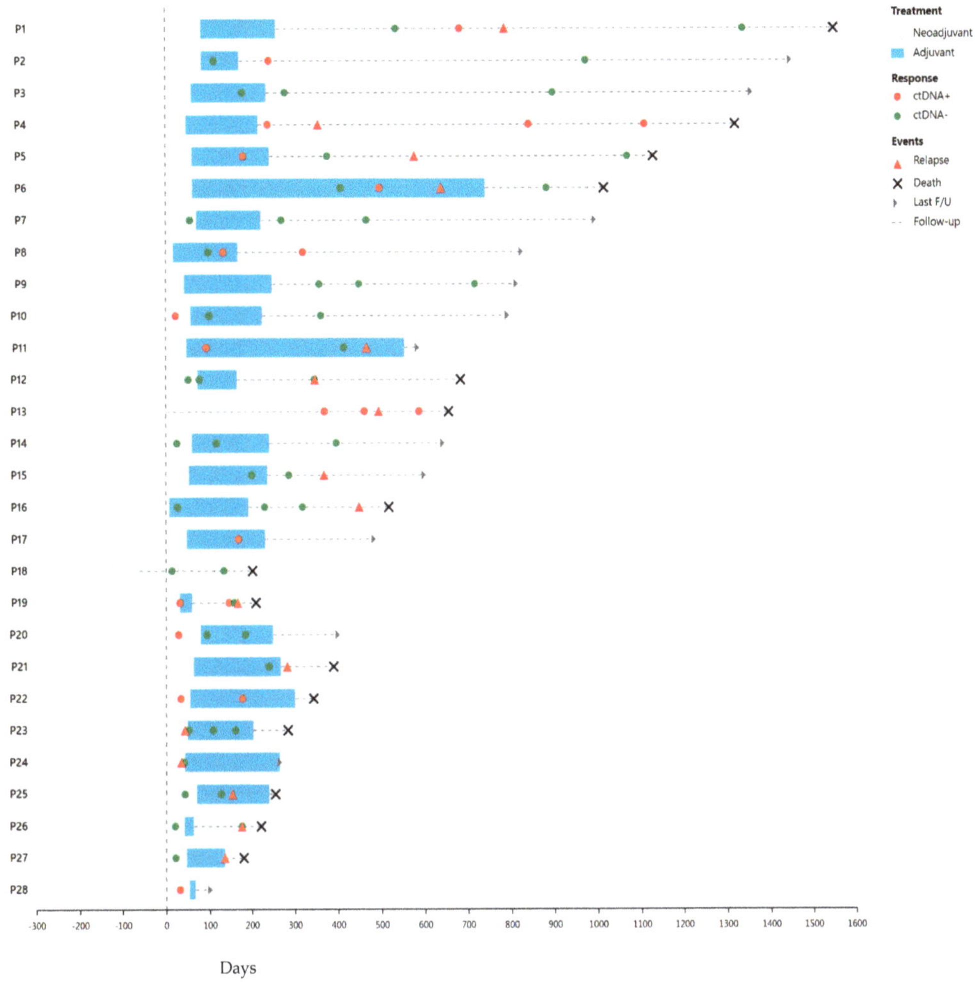

Figure 1. ctDNA for MRD detection after surgery. Patients are monitored before and after surgery (on Day 0) for relapse as confirmed on imaging with ctDNA and followed until death, data cut off time, or being lost to follow up.

3.3. *ctDNA as a Prognostic Marker*

Of fifteen patients with ctDNA collected within the MRD window after surgery, eight had positive ctDNA while seven had negative ctDNA. Fourteen of the fifteen (including all but one ctDNA (−) patient who had received neoadjuvant chemotherapy) received adjuvant chemotherapy. Of patients who underwent resection for PDAC, 14 had at least one instance of ctDNA positivity associated with disease relapses or progression, with a median time between positive ctDNA and CT-confirmed relapse/progression of 81 days (Figure 1). This highlights the utility of ctDNA in predicting disease progression.

Median follow up was significantly different between ctDNA (−) and ctDNA (+) groups, at 22 and 11 months, respectively (p = 0.009). In those with initially negative ctDNA (including two with positive surgical margins), seven had progression after a median of 12.3 months (range 5.5–37.4 months), compared to mPFS of 8.7 months (range 1.1–26.2 months) in ctDNA (+) patients. Most patients who received adjuvant chemotherapy progressed while on treatment (see Figure 1). The median overall survival (OS) for ctDNA (−) patients was significantly longer at 37.6 (95% CI: 20.9–49.5) months, compared to ctDNA (+) patients (13.4 (95% CI: 6–21.8) months) (see Figure 2). Overall, a positive initial ctDNA is associated with higher mortality compared to a negative ctDNA test (HR: 3.97, 95% CI: 1.61–9.80, $p = 0.003$).

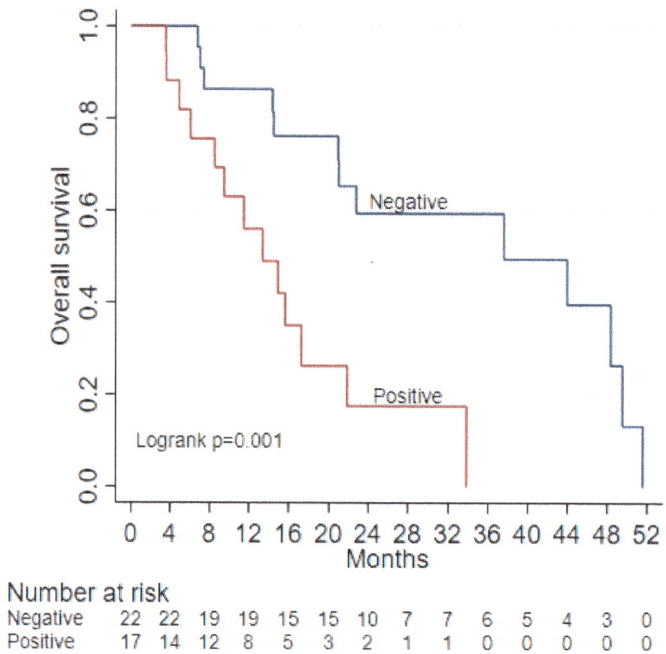

Figure 2. Overall survival by initial ctDNA.

4. Discussion

Previous studies have shown that positive ctDNA is associated with shorter PFS and OS in patients who undergo resection for pancreatic cancer [26–28]. For example, one study reported a median survival of 13.6 months vs. 27.6 months in patients with detectable vs. no detectable ctDNA, which is similar to the findings from our study [35]. Compared to other surveillance tools (e.g., CA19-9 and ctDNA detection using a panel of commonly mutated genes), tumor-informed ctDNA has been proposed to be a more sensitive and specific detection method with a high prognostic value [27,28]. In this study, using tumor-informed ctDNA samples from patients at our institution, we found a 24.2-month difference in OS between ctDNA (−) and ctDNA (+) patients, which further supports the use of ctDNA as a prognostic biomarker in PDAC.

As adjuvant chemotherapy is indicated for localized pancreatic cancer, tumor-informed ctDNA can be collected within the MRD window as a prognostic tool and beyond the MRD window to monitor for recurrence. In a meta-analysis by Lee et al. [37], positive ctDNA at baseline and postoperatively is associated with decreased overall survival (HR 2.27 and HR 3.66, respectively) and a higher risk of disease recurrence [37]. In the current study, patients were tested for ctDNA within and beyond the MRD window. Positive ctDNA was not associated with age, sex, or location of the pancreatic tumor. However, a lack of ctDNA detection was associated with negative margins and the lack of lymphovascular invasion (LVI), which are, in themselves, positive prognostic factors. Given the higher percentage of high-risk features in our study—positive margins (36% vs. 5%) and lymphovascular invasion (91% vs. 32%)—in patients with initial ctDNA+ disease, adjuvant therapy may be especially warranted for this patient population. Indeed, data from this as well as other studies supports ctDNA testing in tandem with other prognostic factors to predict patient outcomes and guide management decisions.

Monitoring for recurrence is uniquely challenging for PDAC compared to other gastrointestinal malignancies such as colorectal carcinoma and cholangiocarcinoma given the lack of sensitive and specific tumor markers. Given the more aggressive nature of the disease, a more sensitive yet accessible test has been long sought-after. As a surveillance test, we show that ctDNA has a higher sensitivity than the more commonly used CA19-9. Sensitivity and specificity increase when both are used to monitor biochemical disease progression even before detection by imaging. Considering this, ctDNA alone or in conjunction with CA19-9 can be used to more reliably track treatment response. A previous study demonstrated that ctDNA levels decreased significantly after treatment initiation but increased at the time of progression, providing a lead time over CA19-9 [38]. As such, ctDNA may be a powerful tool to guide treatment decisions in all phases of PDAC including localized, borderline-resectable, and metastatic disease. Persistent ctDNA positivity indicates aggressive disease biology that may require alternative therapies, while a negative result could lead to treatment de-escalation.

The study has a few considerations, including the cohort size and its retrospective, single-institutional design. While only a subset of patients underwent ctDNA testing within the MRD window, the findings suggest that a positive ctDNA result, regardless of when it is found with regard to diagnosis or surgery, is strongly associated with a worse prognosis. This can help guide clinicians toward closer surveillance and more aggressive treatment options for high-risk patients. The absence of preoperative ctDNA samples in resected patients is another consideration, though future studies may explore the prognostic value of changes in ctDNA levels postoperatively. Despite these factors, this study provides valuable insights into ctDNA's role in PDAC management, supporting its potential as a robust biomarker for disease monitoring and prognosis.

Randomized prospective trials are needed to assess the value of ctDNA in the prognosis and risk stratification for neoadjuvant and adjuvant management in PDAC. Currently, the CASPER trial (NCT05853198) seeks to evaluate ctDNA as a marker of surgical futility in a single arm of patients with operable PDAC, with an endpoint of relapse within 2 years of surgery. In addition, the FRENCH.MRD.PDAC is a French trial yet to begin recruitment, that aims to determine disease-free survival (DFS) and ctDNA detection status after curative surgery and adjuvant chemotherapy. Overall, ctDNA promises to be a valuable tool for personalized management of PDAC patients.

Author Contributions: M.A. and A.E. conceptualized and designed this retrospective study, with A.E. also providing administrative support. M.A., A.E., Y.Z., H.H., V.D., W.A., E.A.-N. and B.K. were responsible for providing study materials and patient-related data. The collection and assembly of data were handled by B.K. and E.A.-N., while Y.Z. and M.A. led the analysis and interpretation efforts. All authors—Y.Z., A.E., H.H., V.D., W.A., E.A.-N., B.K. and M.A.—actively participated in writing the manuscript for this study and gave their final approval to the completed document. All authors have read and agreed to the published version of the manuscript.

Funding: This research received no external funding.

Institutional Review Board Statement: The authors are accountable for all aspects of the work in ensuring that questions related to the accuracy or integrity of any part of the work are appropriately investigated and resolved. The study was conducted in accordance with the Declaration of Helsinki (as revised in 2013). The study was approved by the Institutional Review Board of Houston Methodist Hospital (No. PRO00035386), 25 July 2023.

Informed Consent Statement: Informed consent was waived due to the use of de-identified data collected as part of routine clinical care.

Data Availability Statement: The data from this study that support our results are available upon request from the corresponding author, Maen Abdelrahim.

Acknowledgments: We would like to express our deepest appreciation and gratitude to The Cockrell Center for Advanced Therapeutics, The William and Ella Owens Medical Research Foundation, and the Houston Methodist Hospital Foundation for their support.

Conflicts of Interest: The authors declare no conflicts of interest.

References

1. Siegel, R.L.; Miller, K.D.; Jemal, A. Cancer statistics, 2019. *CA Cancer J. Clin.* **2019**, *69*, 7–34. [CrossRef]
2. Abdelrahim, M.; Esmail, A.; Kasi, A.; Esnaola, N.F.; Xiu, J.; Baca, Y.; Weinberg, B.A. Comparative molecular profiling of pancreatic ductal adenocarcinoma of the head versus body and tail. *NPJ Precis. Oncol.* **2024**, *8*, 85. [CrossRef] [PubMed]
3. Bugazia, D.; Al-Najjar, E.; Esmail, A.; Abdelrahim, S.; Abboud, K.; Abdelrahim, A.; Umoru, G.; Rayyan, H.A.; Abudayyeh, A.; Al Moustafa, A.E.; et al. Pancreatic ductal adenocarcinoma: The latest on diagnosis, molecular profiling, and systemic treatments. *Front. Oncol.* **2024**, *14*, 1386699. [CrossRef]
4. *Pancreatic Cancer Statistics*; American Cancer Society: Atlanta, GA, USA, 2024.
5. Rahib, L.; Smith, B.D.; Aizenberg, R.; Rosenzweig, A.B.; Fleshman, J.M.; Matrisian, L.M. Projecting cancer incidence and deaths to 2030: The unexpected burden of thyroid, liver, and pancreas cancers in the United States. *Cancer Res.* **2014**, *74*, 2913–2921. [CrossRef]
6. Ducreux, M.; Cuhna, A.S.; Caramella, C.; Hollebecque, A.; Burtin, P.; Goéré, D.; Seufferlein, T.; Haustermans, K.; Van Laethem, J.L.; Conroy, T.; et al. Cancer of the pancreas: ESMO Clinical Practice Guidelines for diagnosis, treatment and follow-up. *Ann. Oncol.* **2015**, *26* (Suppl. S5), v56–v68. [CrossRef] [PubMed]
7. Badheeb, M.; Abdelrahim, A.; Esmail, A.; Umoru, G.; Abboud, K.; Al-Najjar, E.; Rasheed, G.; Alkhulaifawi, M.; Abudayyeh, A.; Abdelrahim, M. Pancreatic tumorigenesis: Precursors, genetic risk factors and screening. *Curr. Oncol.* **2022**, *29*, 8693–8719. [CrossRef] [PubMed]
8. Esmail, A.; Badheeb, M.; Abdelrahim, M. Pancreatic Tumorigenesis: Precursors, Genetic Risk Factors and Screening. In *Pancreatic Cancer-Updates in Pathogenesis, Diagnosis and Therapies*; IntechOpen: London, UK, 2023.
9. Adamska, A.; Domenichini, A.; Falasca, M. Pancreatic Ductal Adenocarcinoma: Current and Evolving Therapies. *Int. J. Mol. Sci.* **2017**, *18*, 1338. [CrossRef]
10. Oettle, H.; Neuhaus, P.; Hochhaus, A.; Hartmann, J.T.; Gellert, K.; Ridwelski, K.; Niedergethmann, M.; Zülke, C.; Fahlke, J.; Arning, M.B.; et al. Adjuvant chemotherapy with gemcitabine and long-term outcomes among patients with resected pancreatic cancer: The CONKO-001 randomized trial. *JAMA* **2013**, *310*, 1473–1481. [CrossRef]
11. Kwaśniewska, D.; Fudalej, M.; Nurzyński, P.; Badowska-Kozakiewicz, A.; Czerw, A.; Cipora, E.; Sygit, K.; Bandurska, E.; Deptała, A. How A Patient with Resectable or Borderline Resectable Pancreatic Cancer should Be Treated-A Comprehensive Review. *Cancers* **2023**, *15*, 4275. [CrossRef]
12. Chikhladze, S.; Lederer, A.K.; Kousoulas, L.; Reinmuth, M.; Sick, O.; Fichtner-Feigl, S.; Wittel, U.A. Adjuvant chemotherapy after surgery for pancreatic ductal adenocarcinoma: Retrospective real-life data. *World J. Surg. Oncol.* **2019**, *17*, 185. [CrossRef]
13. Lee, J.C.; Ahn, S.; Cho, I.K.; Lee, J.; Kim, J.; Hwang, J.H. Management of recurrent pancreatic cancer after surgical resection: A protocol for systematic review, evidence mapping and meta-analysis. *BMJ Open* **2018**, *8*, e017249. [CrossRef]
14. Park, W.; Chawla, A.; O'Reilly, E.M. Pancreatic Cancer: A Review. *JAMA* **2021**, *326*, 851–862. [CrossRef] [PubMed]
15. Tempero, M.A.; Malafa, M.P.; Al-Hawary, M.; Behrman, S.W.; Benson, A.B.; Cardin, D.B.; Chiorean, E.G.; Chung, V.; Czito, B.; Del Chiaro, M.; et al. Pancreatic Adenocarcinoma, Version 2.2021, NCCN Clinical Practice Guidelines in Oncology. *J. Natl. Compr. Canc. Netw.* **2021**, *19*, 439–457. [CrossRef] [PubMed]
16. Aoki, T.; Mori, S.; Kubota, K. Neoadjuvant and Adjuvant Chemotherapy for Pancreatic Adenocarcinoma: Literature Review and Our Experience of NAC-GS. *Cancers* **2024**, *16*, 910. [CrossRef] [PubMed]
17. Vivarelli, M.; Mocchegiani, F.; Nicolini, D.; Vecchi, A.; Conte, G.; Dalla Bona, E.; Rossi, R.; Benedetti Cacciaguerra, A. Neoadjuvant Treatment in Resectable Pancreatic Cancer. Is It Time for Pushing on It? *Front. Oncol.* **2022**, *12*, 914203. [CrossRef]

18. Sugumar, K.; Hue, J.J.; De La Serna, S.; Rothermel, L.D.; Ocuin, L.M.; Hardacre, J.M.; Ammori, J.B.; Winter, J.M. The importance of time-to-adjuvant treatment on survival with pancreatic cancer: A systematic review and meta-analysis. *Cancer Rep.* **2021**, *4*, e1390. [CrossRef]
19. Khorana, A.A.; McKernin, S.E.; Berlin, J.; Hong, T.S.; Maitra, A.; Moravek, C.; Mumber, M.; Schulick, R.; Zeh, H.J.; Katz, M.H.G. Potentially Curable Pancreatic Adenocarcinoma: ASCO Clinical Practice Guideline Update. *J. Clin. Oncol.* **2019**, *37*, 2082–2088. [CrossRef]
20. Smaglo, B.G. Role for Neoadjuvant Systemic Therapy for Potentially Resectable Pancreatic Cancer. *Cancers* **2023**, *15*, 2377. [CrossRef]
21. Lee, T.; Teng, T.Z.J.; Shelat, V.G. Carbohydrate antigen 19-9—Tumor marker: Past, present, and future. *World J. Gastrointest. Surg.* **2020**, *12*, 468–490. [CrossRef]
22. Kim, S.; Park, B.K.; Seo, J.H.; Choi, J.; Choi, J.W.; Lee, C.K.; Chung, J.B.; Park, Y.; Kim, D.W. Carbohydrate antigen 19-9 elevation without evidence of malignant or pancreatobiliary diseases. *Sci. Rep.* **2020**, *10*, 8820. [CrossRef]
23. Tsen, A.; Barbara, M.; Rosenkranz, L. Dilemma of elevated CA 19-9 in biliary pathology. *Pancreatology* **2018**, *18*, 862–867. [CrossRef] [PubMed]
24. Campos, N.M.F.; Almeida, V.; Curvo Semedo, L. Peritoneal disease: Key imaging findings that help in the differential diagnosis. *Br. J. Radiol.* **2022**, *95*, 20210346. [CrossRef]
25. Dong, Q.; Chen, C.; Hu, Y.; Zhang, W.; Yang, X.; Qi, Y.; Zhu, C.; Chen, X.; Shen, X.; Ji, W. Clinical application of molecular residual disease detection by circulation tumor DNA in solid cancers and a comparison of technologies: Review article. *Cancer Biol. Ther.* **2023**, *24*, 2274123. [CrossRef]
26. Watanabe, F.; Suzuki, K.; Aizawa, H.; Endo, Y.; Takayama, Y.; Kakizawa, N.; Kato, T.; Noda, H.; Rikiyama, T. Circulating tumor DNA in molecular assessment feasibly predicts early progression of pancreatic cancer that cannot be identified via initial imaging. *Sci. Rep.* **2023**, *13*, 4809. [CrossRef]
27. Botta, G.P.; Abdelrahim, M.; Aushev, V.N.; Esmail, A.; Drummond, B.; Sharma, S.; Kalashnikova, E.; Hook, N.; Chandana, S.R.; Tejani, M.A.; et al. Association of personalized and tumor-informed ctDNA with patient survival outcomes in pancreatic adenocarcinoma. *J. Clin. Oncol.* **2022**, *40*, 517. [CrossRef]
28. Botta, G.P.; Abdelrahim, M.; Drengler, R.L.; Aushev, V.N.; Esmail, A.; Laliotis, G.; Brewer, C.M.; George, G.V.; Abbate, S.M.; Chandana, S.R.; et al. Association of personalized and tumor-informed ctDNA with patient survival outcomes in pancreatic adenocarcinoma. *Oncologist* **2024**, *29*, 859–869. [CrossRef] [PubMed]
29. Marrugo-Ramírez, J.; Mir, M.; Samitier, J. Blood-Based Cancer Biomarkers in Liquid Biopsy: A Promising Non-Invasive Alternative to Tissue Biopsy. *Int. J. Mol. Sci.* **2018**, *19*, 2877. [CrossRef]
30. Crowley, E.; Di Nicolantonio, F.; Loupakis, F.; Bardelli, A. Liquid biopsy: Monitoring cancer-genetics in the blood. *Nat. Rev. Clin. Oncol.* **2013**, *10*, 472–484. [CrossRef] [PubMed]
31. Vymetalkova, V.; Cervena, K.; Bartu, L.; Vodicka, P. Circulating Cell-Free DNA and Colorectal Cancer: A Systematic Review. *Int. J. Mol. Sci.* **2018**, *19*, 3356. [CrossRef]
32. Merker, J.D.; Oxnard, G.R.; Compton, C.; Diehn, M.; Hurley, P.; Lazar, A.J.; Lindeman, N.; Lockwood, C.M.; Rai, A.J.; Schilsky, R.L.; et al. Circulating Tumor DNA Analysis in Patients With Cancer: American Society of Clinical Oncology and College of American Pathologists Joint Review. *J. Clin. Oncol.* **2018**, *36*, 1631–1641. [CrossRef]
33. Samandari, M.; Julia, M.G.; Rice, A.; Chronopoulos, A.; Del Rio Hernandez, A.E. Liquid biopsies for management of pancreatic cancer. *Transl. Res. J. Lab. Clin. Med.* **2018**, *201*, 98–127. [CrossRef] [PubMed]
34. Patel, H.; Okamura, R.; Fanta, P.; Patel, C.; Lanman, R.B.; Raymond, V.M.; Kato, S.; Kurzrock, R. Clinical correlates of blood-derived circulating tumor DNA in pancreatic cancer. *J. Hematol. Oncol.* **2019**, *12*, 130. [CrossRef] [PubMed]
35. Hadano, N.; Murakami, Y.; Uemura, K.; Hashimoto, Y.; Kondo, N.; Nakagawa, N.; Sueda, T.; Hiyama, E. Prognostic value of circulating tumour DNA in patients undergoing curative resection for pancreatic cancer. *Br. J. Cancer* **2016**, *115*, 59–65. [CrossRef]
36. Grunvald, M.W.; Jacobson, R.A.; Kuzel, T.M.; Pappas, S.G.; Masood, A. Current Status of Circulating Tumor DNA Liquid Biopsy in Pancreatic Cancer. *Int. J. Mol. Sci.* **2020**, *21*, 7651. [CrossRef] [PubMed]
37. Lee, J.S.; Rhee, T.M.; Pietrasz, D.; Bachet, J.B.; Laurent-Puig, P.; Kong, S.Y.; Takai, E.; Yachida, S.; Shibata, T.; Lee, J.W.; et al. Circulating tumor DNA as a prognostic indicator in resectable pancreatic ductal adenocarcinoma: A systematic review and meta-analysis. *Sci. Rep.* **2019**, *9*, 16971. [CrossRef]
38. Lapin, M.; Edland, K.H.; Tjensvoll, K.; Oltedal, S.; Austdal, M.; Garresori, H.; Rozenholc, Y.; Gilje, B.; Nordgård, O. Comprehensive ctDNA Measurements Improve Prediction of Clinical Outcomes and Enable Dynamic Tracking of Disease Progression in Advanced Pancreatic Cancer. *Clin. Cancer Res.* **2023**, *29*, 1267–1278. [CrossRef]

biomedicines

Article

Safety and Efficacy of Regorafenib and 5-Fluorouracil Combination Therapy in Refractory Metastatic Colorectal Cancer After Third-Line Treatment: An Institutional Experience

Maen Abdelrahim [1,*], Abdullah Esmail [1], Ebtesam Al-Najjar [1], Bayan Khasawneh [1], Godsfavour Umoru [1], Waseem Abdelrahim [2], Karen Abboud [1] and Veronica B. Ajewole [1]

1 Section of GI Oncology, Houston Methodist Neal Cancer Center, Houston Methodist Hospital, Houston, TX 77094, USA; aesmail@houstonmethodist.org (A.E.)
2 Michael E. DeBakey HS for Health Professions, Houston, TX 77030, USA
* Correspondence: mabdelrahim@houstonmethodist.org

Abstract: Background: Colorectal carcinoma (CRC) is one of the most common cancer types along with breast, prostate, and lung cancer. Many patients with CRC present with metastatic disease despite receiving standard first- and second-line therapies; thus emerges the demand for implementing new therapies that could improve outcomes among CRC patients. This case series was conducted to assess the efficacy and safety of regorafenib plus 5-fluorouracil (5-FU) in patients with refractory metastatic CRC (mCRC). **Methods**: We conducted a retrospective analysis of data from adult patients aged 18 and above who were diagnosed with refractory mCRC and received regorafenib plus 5-FU combination therapy at Houston Methodist Hospital between November 2017 and October 2023. Our study focuses on assessing key outcomes, including Overall Survival [OS], Progression-Free Survival [PFS], and safety. **Results**: Among the 12 patients we included in this study who underwent regorafenib plus 5-FU combination therapy for refractory mCRC after receiving at least three prior lines of treatment, the best response for six patients (50%) was successfully achieved, with disease control within 7–12 weeks from therapy initiation. Patients had an overall good tolerance for this treatment regimen and reported only the most common adverse events, including Hand-Foot Syndrome (HFS), mucositis, and hypertension (HTN), which were mostly resolved with dose adjustment of medications. **Conclusions**: This study highlights that using a combination of regorafenib plus 5-FU can be a potential treatment option for patients with refractory mCRC. Additional research, including prospective clinical trials, is required to assess the effectiveness and safety of regorafenib and 5-FU combination therapy in comparison to other currently limited treatment options.

Keywords: regorafenib plus 5-FU; 5-fluorouracil; refractory mCRC; multi-kinase inhibitor; OS

1. Introduction

Colorectal carcinoma (CRC) is one of the most diagnosed cancers around the world. It affects both men and women, but the numbers vary slightly between genders [1]. For men, CRC is the third most prevalent cancer after lung and prostate cancers. This means that out of all the men diagnosed with cancer, about 10% of them have CRC. That is approximately 746,000 cases. Among women, CRC is the second most prevalent cancer after breast cancer. Thus, about 9.2% of all women diagnosed with cancer have CRC. That is approximately 614,000 cases [1].

In the United States, CRC is the third most diagnosed and the third leading cause of cancer-related deaths for both men and women. However, it ranks second in cancer-related deaths overall and is the leading cause of death among men younger than 50 years. More than half of all cases and deaths can be linked to modifiable risk factors including smoking, unhealthy diet, high alcohol consumption, physical inactivity, and excess body weight. CRC stands as the second most prevalent cause of cancer-related deaths in the United States. Every three years, the American Cancer Society (ACS) releases updated statistics on CRC, extracting data on incidence from population-based cancer registries and mortality data from the National Center for Health Statistics (NCHS). In 2023, it is projected that approximately 153,020 individuals will be diagnosed with CRC, while 52,550 will die from the disease. Among those cases, 19,550 will affect individuals under the age of 50, resulting in 3750 deaths [2]. Among people diagnosed with mCRC, approximately 70% to 75% of patients survive beyond 1 year, 30% to 35% beyond 3 years, and fewer than 20% beyond 5 years from diagnosis. The treatment strategy for mCRC is termed resectable when the primary tumor and all metastases are amenable to complete surgical removal. However, in these patients, nodal infiltration and occult micro-metastatic dissemination are common. Resection of mCRC achieves long-term cure for less than 20% of mCRC patients [3]. Currently, several different drugs are being used in mCRC treatment, with first- and second-line options including the systemic drugs 5-FU, irinotecan, oxaliplatin, bevacizumab, cetuximab, panitumumab, and the oral drug capecitabine, as well as different combinations of these drugs, such as the FOLFOX regimen (5-FU, leucovorin, and oxaliplatin), the FOLFIRI regimen (5-FU, leucovorin, and irinotecan), and the XELOX regimen (capecitabine and oxaliplatin), either with or without a monoclonal antibody agent [4]. However, treatment options beyond the third line of refractory mCRC remain challenging, and despite the availability of multiple options, outcomes are generally poor [5].

Regorafenib is an orally active, potent multi-kinase inhibitor that targets a broad range of angiogenic, stromal, and oncogenic kinases, including Vascular Endothelial Growth Factor (VEGF) receptors 1, 2, and 3, tyrosine kinase with immunoglobulin and epidermal growth factor homology domain 2 (TIE-2), platelet-derived growth factor receptor-β, c-kit, ret, raf-1, and BRAF [6,7].

5-FU is an antimetabolite drug that has been widely used since 1957 to treat different types of cancer, including CRC and breast cancer [8]. 5-FU is a pyrimidine analogue that acts as an antimetabolite of uracil, causing cell death. After entering the cell, 5-FU is converted to one of several active metabolites. These metabolites, in turn, inhibit the enzymes thymidylate synthase and uracil-DNA-glycosylase, interfering with DNA synthesis and repair, respectively. In addition, one of the 5-FU metabolites is incorporated into RNA, thereby disrupting its processing and function [9].

In September 2012, the US Food and Drug Administration (FDA) approved regorafenib for the treatment of mCRC patients who had failed FOLFIRI chemotherapy regimens; an anti-VEGF pathway therapy; and an anti-Epidermal Growth Factor Receptor (EGFR) therapy (for KRAS wild-type patients). In the phase III CORRECT (Regorafenib Monotherapy for Previously Treated mCRC) trial data demonstrated improved OS benefit for mCRC patients treated with regorafenib versus placebo in patients with treatment-refractory mCRC (6.4 vs. 5.0 months; $p = 0.0052$) [10].

In recent years, numerous studies and hypotheses have been explored to combine regorafenib with various drugs, aiming to establish and validate its potential synergistic effects. These investigations have been driven by the pursuit of enhanced therapeutic outcomes, particularly in the context of treatment-refractory mCRC. One of these hypotheses is that combining regorafenib with anti-programmed cell death 1 (PD-1)/anti-programmed cell death ligand 1 (PD-L1), antibodies may be associated with significant clinical benefit in patients with mCRC. The median PFS and OS were 3.6 months [95% confidence interval (CI), 1.8–5.4] and 10.8 months (95% CI, 5.9–NA), respectively [11].

In other retrospectively analyzed patients with advanced or mCRC who received at least one dose of immune checkpoint inhibitors ICIs combined with regorafenib in 14 Chinese medical centers, the median PFS was 3.1 months (95% CI, 2.3–4.2) and the median OS was 17.3 months [12].

While prior studies explored regorafenib with fluoropyrimidines, this study uniquely provides real-world dosing optimization and outcomes in a rare, off-label refractory mCRC population, addressing a critical gap in practical guidance for heavily pretreated patients. This study aims to evaluate the efficacy and safety of regorafenib combined with intravenous 5-FU in heavily pretreated patients with mCRC who have progressed beyond third-line therapy. Additionally, it seeks to provide practical insight on dosing and toxicity management for this off-label combination, addressing a critical gap in real-world data.

2. Materials and Methods

2.1. Study Design and Participants

This case series evaluates the treatment outcomes for a total of 12 mCRC patients. The inclusion criteria for the study were pathology-confirmed adenocarcinoma of the colon or rectum with radiological evidence of mCRC, age of at least 18 years, patients who had received at least three lines of previous treatment for mCRC, and Eastern Cooperative Oncology Group Performance Status (ECOG PS) score restricted to 0–1. Data were collected for patients treated at Houston Methodist Cancer Center between November 2017 and October 2023. The study cohort of 12 patients (2017–2023) includes 7 patients previously reported in Haque et al. (2017–2021) with extended follow-up, plus 5 additional patients enrolled subsequently [5].

We collected demographic information, including age, gender, ECOG PS, primary site of CRC, metastatic sites (liver and lung), primary tumor surgical interventions and metastasectomies, prior adjuvant chemotherapy, number of previous chemotherapy sessions, initial dose of regorafenib, and presence of KRAS mutations. We gathered information on adverse events associated with regorafenib, including Hand-Foot Syndrome (HFS), Hypertension (HTN), skin rash, and instances of emergency hospitalization. Adverse event severity was assessed based on the National Cancer Institute Common Terminology Criteria for Adverse Events (NCI-CTCAE). Good tolerance was defined as completing at least two treatment cycles without grade 3/4 toxicities requiring permanent discontinuation, maintaining a regorafenib dose of at least 80 mg/day, and avoiding hospitalization due to treatment-related adverse events. These data were retrospectively gathered from Electronic Medical Records (EMR). Approval for this study was granted by the Institutional Review Board (IRB) of the Houston Methodist Research Institute.

2.2. Patients' Characteristics

A total of 12 patients with mCRC were included in this study, including eight males and four females, with an average age of 59 years old. Ethnically, seven of the patients were of Caucasian race, four were of Asian race, and one was of black race. Patients' age ranged from 31–60 years old at the time of diagnosis and from 40–65 years old at therapy initiation. Most of the enrolled patients had HTN as a co-morbid condition. Seven patients (58%) had left-sided CRC, two patients (16%) had right-sided CRC, and three patients (25%) had rectal cancer. The primary tumor was resected in eleven patients (91%). Nine patients (75%) had stage IV at the time of diagnosis, two patients (16%) had stage III, and one patient had stage I. All patients had mutated KRAS or NRAS except for one patient and none had microsatellite instability.

The prevalent metastatic site was confirmed through imaging such as Computed Tomography (CT) and Magnetic Resonance Imaging (MRI) scans. The liver was identified as the primary location (in all 12 patients) followed by the lung (in 8 patients). All patients received (FOLFOX) chemotherapy and anti-VEGF therapy prior to combination therapy; three patients started with FOLFOX, two with capecitabine and oxaliplatin, and one with capecitabine monotherapy. The second-line chemotherapy for most patients comprised (FOLFIRI) and bevacizumab. Additionally, two patients received regorafenib as monotherapy prior to the use of combination therapy. Other commonly used therapies include trifluridine/tipiracil, ramucirumab, and capecitabine. The information on therapies used prior to regorafenib plus 5-FU combination therapy is listed in Table 1.

2.3. Treatment and Assessment

The clinical outcomes of interest encompassed the optimal response to the combined therapeutic regimen involving regorafenib and 5-FU. Safety data were available for the twelve patients who received each regimen.

Regorafenib is typically prescribed at a standard dose of 160 mg (administered as four 40 mg tablets) once a day. This treatment follows a schedule of three weeks on, followed by one week off therapy. Because of the adverse events of regorafenib, the standard doses were not applicable to every patient. Most patients in our study were initiated on a regorafenib dosage of 80 mg/day, a dose that was subsequently escalated within one to two weeks to attain a target dosage of 120 mg. Further adjustments were made, allowing for a maximum dosage of 160 mg, determined by the individual's tolerability.

Regarding 5-FU, dosages were mostly administered as a 400 mg/m^2 bolus, followed by a continuous infusion of 2400 mg/m^2 over 46 h, initiated on the first day of the therapy.

Key endpoints included PFS, OS, and the assessment of adverse events. PFS was defined as the time span between the initiation of combination therapy and the occurrence of clinically substantiated evidence indicating disease progression. On the contrary, OS was described as the timeframe from the initiation of treatment to mortality resulting from any cause.

We kept detailed records of the given doses of 5-FU and regorafenib when treatment started to the point of discontinuation or the time of the last follow-up, as shown in Figure 1. Through this approach, we aimed to evaluate the therapeutic efficacy, survival outcomes, and safety with combined treatment administration.

Table 1. Basic characteristics of patients.

Patient	Age at Therapy Initiation (Gender)	Diagnosis (Stage at Diagnosis)	Chemotherapies Prior to Rego + 5-FU (Including Maintenance Therapy)	History of Tumor Resection	Mutations	Sites of Metastasis	Comorbidities
1	65 (Female)	Left-sided colon cancer. (IV)	FOLFOX + bevacizumab FOLFIRI + bevacizumab bevacizumab monotherapy Trifluridine/Tipiracil initially with bevacizumab and then Ramucirumab FOLFIRINOX + bevacizumab	Yes	KRAS-mutant (G12D) NRAS-neg BRAF-neg MSI-stable	Peritoneum, ovaries, abdominal wall, liver	Osteoarthritis
2	48 (Male)	Left-sided colon cancer. (III)	Capecitabine + Oxaliplatin Capecitabine monotherapy FOLFIRI + bevacizumab FOLFOX + Oxaliplatin	Yes	KRAS-mutated MYC-mutated TP53-mutated NTRK 1–3-neg	Liver	Hypertension
3	58 (Male)	Right-sided colon cancer (IV)	FOLFOX FOLFIRI + bevacizumab Restarted FOLFIRI + bevacizumab.	Yes	KRAS-mutated TP53-mutated PIK3C-mutated BRAF-neg NTRK 1–3-neg MSI-stable HER-2-neg NRAS-neg	Liver and lung	Hypertension Diabetes Mellitus
4	46 (Male)	Left-sided colon cancer. (II)	FOLFOX Capecitabine + oxaliplatin + bevacizumab Capecitabine + bevacizumab bevacizumab FOLFOX + bevacizumab bevacizumab + 5-FU FOLFOX + bevacizumab FOLFIRI + bevacizumab Trifluridine/Tipiracil	Yes	NRAS mutation MSI-stable, FAP with pathogenic APC variant. HER2 negative.	Liver, lungs, bone, and peritoneum	None
5	63 (Female)	Left-sided colon cancer (IV)	FOLFOX + bevacizumab FOLFIRI + ramucirumab FOLFIRI + aflibercept Trifluridine/Tipiracil 5-FU	Yes	NRAS-mutated KRAS-neg BRAF-neg MSI-Stable	Liver and lung	Hypothyroidism Nephropathy

Table 1. *Cont.*

Patient	Age at Therapy Initiation (Gender)	Diagnosis (Stage at Diagnosis)	Chemotherapies Prior to Rego + 5-FU (Including Maintenance Therapy)	History of Tumor Resection	Mutations	Sites of Metastasis	Comorbidities
6	58 (Male)	Left-sided colon cancer (IV)	FOLFOX FOLFIRI + bevacizumab Trifluridine/Tipiracil Capecitabine + oxaliplatin + bevacizumab	Yes	-BRAF-neg -KRAS-mutated -NRAS-neg -HRAS-neg -MSI-Stable	Liver and lung	Hypertension Hypercholesterolemia
7	59 (Male)	Rectal cancer (IV)	Capecitabine FOLFOX capecitabine + bevacizumab irinotecan + bevacizumab bevacizumab panitumumab + irinotecan cetuximab + irinotecan 5-FU + irinotecan + cetuximab regorafenib monotherapy	Yes	KRAS wild type MSI stable	Liver and lung	Hypertension Gout
8	46 (Female)	Rectal cancer (IV)	FOLFOX + bevacizumab FOLFIRI + bevacizumab FOLFIRI + cetuximab	No	BRAF D594G pMMR PIK3CA MSI stable	Lung, Liver, Bone, Supraclavicular Lymph nodes	Hypertension
9	51 (Male)	Right-sided colon cancer (IV)	Capecitabine + oxaliplatin FOLFOX + bevacizumab FOLFIRI + bevacizumab capecitabine + bevacizumab bevacizumab + irinotecan + oxaliplatin	Yes	P53 MSS	Liver and Lung	Hypertension
10	52 (Male)	Rectosigmoid cancer (IV)	FOLFOX + bevacizumab FOLFIRI + cetuximab	Yes	FGFR1 MYC MSS	Liver, Lung and Peritoneum	Hypertension
11	60 (Male)	Rectal cancer (IV)	FOLFOX + bevacizumab FOLFIRI + cetuximab	No	HRAS G13C pMMR	Liver and Lymph nodes	Hypertension, Diabetes Mellitus, Asthma, Chronic Kidney Disease, Stroke
12	40 (Female)	Sigmoid cancer (IV)	FOLFOX + bevacizumab FOLFIRI + cetuximab	Yes	PTEN pMMR MSI stable	Liver and Peritoneum	Hypertension

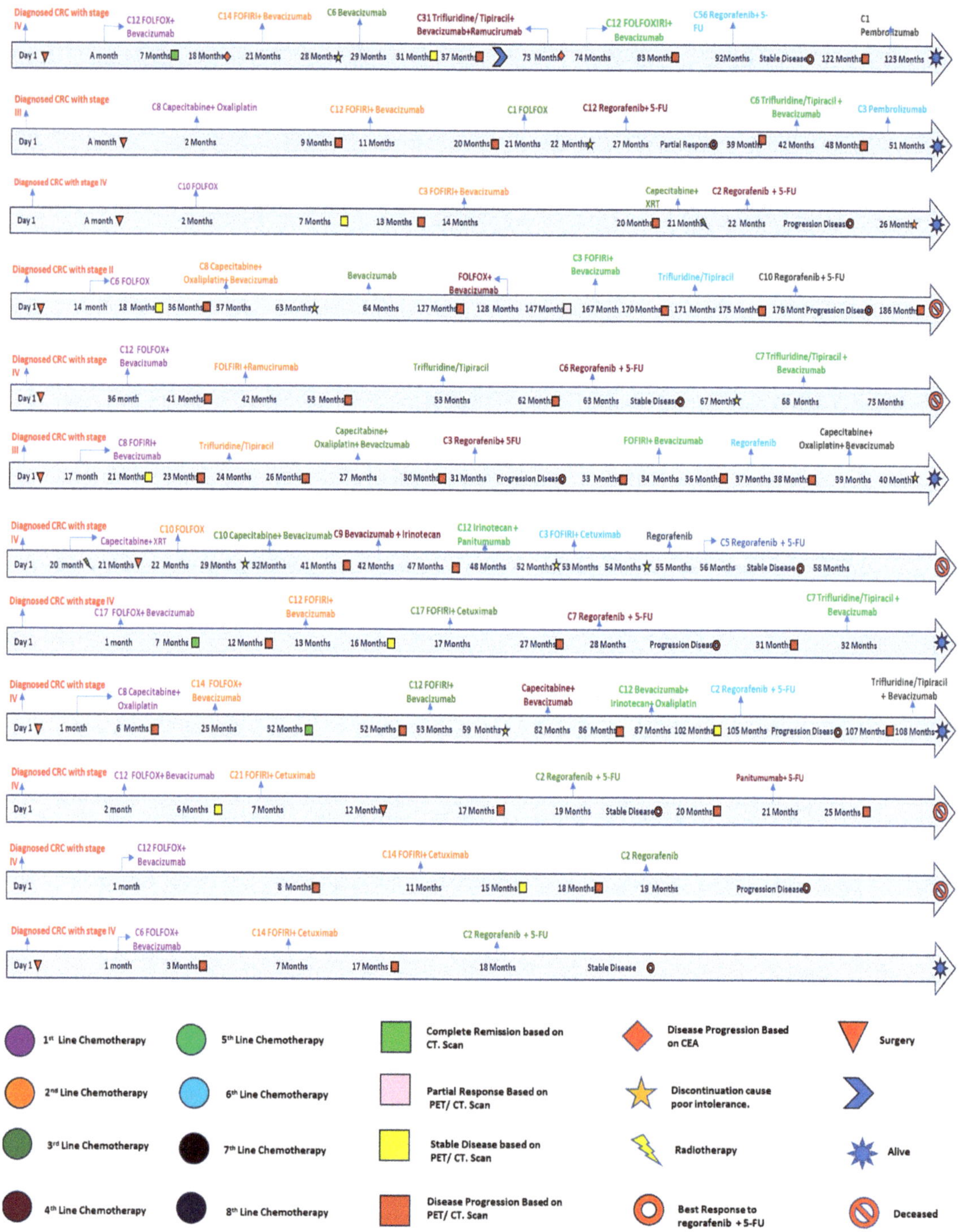

Figure 1. Detailed timelines for regorafenib and 5-FU patients.

3. Results

3.1. Efficacy

Our study yielded promising outcomes among the twelve patients whose tumor responses were evaluated, showcasing disease control (i.e., Partial Response (PR) or Stable Disease (SD)) in six patients. A positive outcome was observed in 50% of patients. Notably, one patient (patient #2) exhibited a PR, while the remaining five patients demonstrated SD status, which were the best responses recorded to this therapy.

Optimal responses were observed within the timeframe of 7–12 weeks after the initiation of the therapeutic regimen. Unfortunately, six patients (50%) exhibited progression of the disease within 6–8 weeks of treatment initiation, requiring a transition to an alternative therapeutic approach. With follow-up duration ranging from 2 to 20 months, seven out of the twelve patients remained alive. A comprehensive summary of the clinical outcomes for all twelve patients is presented in Table 2.

Table 2. Outcomes in patients receiving regorafenib and 5-FU.

Patient #	Best Response (Time to the Best Response From Initiation in Weeks)	Progression or Discontinuation of Therapy (Time to Progression/Therapy Discontinuation)	Therapy after Rego + 5-FU	Time to the Last Follow-Up (in Months)	Status at Last Follow-Up
1	Stable disease (130)	Progression (130)	Pembrolizumab	31	Alive on other therapy
2	Partial response (53)	Progression (53)	Trifluridine/Tipiracil + bevacizumab pembrolizumab	30	Alive on other therapy
3	Progressive disease (7)	Progression (7)	Nivolumab + regorafenib Trifluridine/Tipiracil + bevacizumab Tolfenamic acid	20	Alive on other therapy
4	Progressive disease (8)	Progression (30)	Pembrolizumab + regorafenib	9.6	Deceased
5	Stable disease (17)	Discontinuation—Toxicity (17)	Trifluridine/Tipiracil + bevacizumab	11	Deceased
6	Progressive disease (13)	Progression (13)	FOLFIRI bevacizumab + oxaliplatin + capecitabine	11.5	Alive on other therapy
7	Stable disease (7)	Discontinuation—Toxicity (7)	None	2	Deceased
8	Progressive disease (13)	Progression (13)	Trifluridine/Tipiracil + bevacizumab	4.5	Alive on other therapy
9	Progressive disease (8)	Progression (8)	Trifluridine/Tipiracil + Bevacizumab	8	Alive on other therapy
10	Stable disease (6)	Progression (6.5)	Panitumumab + 5-FU	6	Deceased
11	Progressive disease (4)	Progression (4)	None	1	Deceased
12	Stable disease (7)	Discontinuation (7)	None	5	Alive

In our study, we observed that the median OS is 12 months. The 10 months OS of 61% (95% Confidence Interval (CI):25–84%) and 12 months OS of 50% (95% CI: 18.8–75.3%) OS Kaplan–Meier analysis is illustrated in Figure 2. Additionally, the median PFS was 3.15 months.

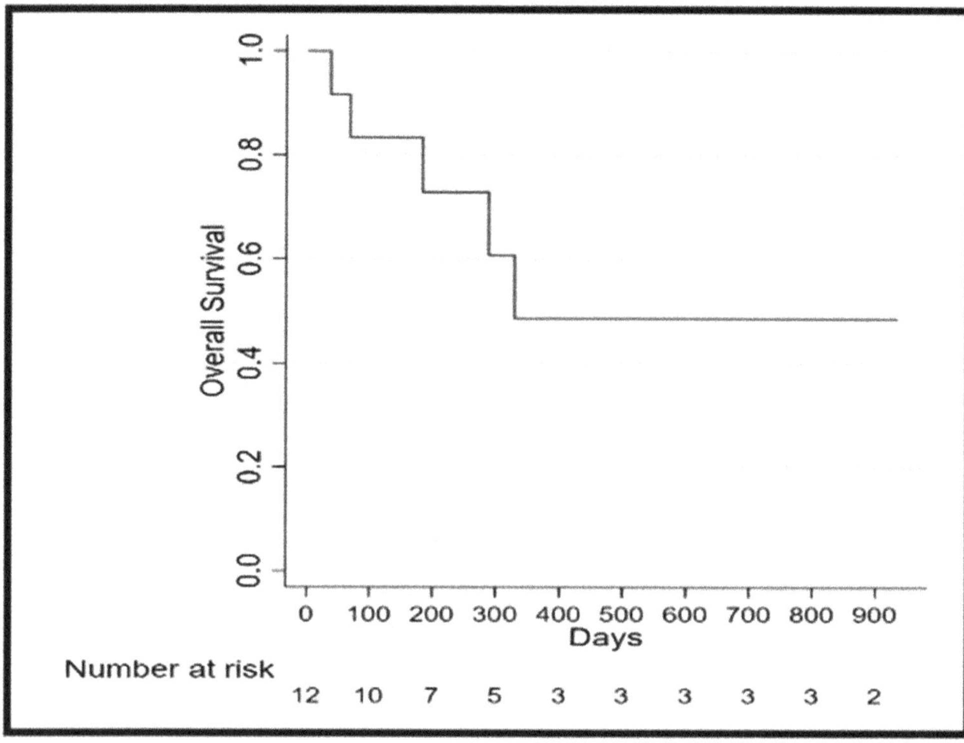

Figure 2. OS Kaplan–Meier analysis.

3.2. Safety

Most of the patients in our study were initiated on a regorafenib dosage of 80 mg/day, a dose that was subsequently escalated within one to two weeks to attain a target dose of 120 mg, with further adjustment to a maximum of 160 mg as deemed tolerable. Notably, six patients (50%) initiated therapy with a dose of 80 mg, five (41%) received an initial dose of 120 mg, and one patient (8%) embarked on treatment with a dose of 160 mg. Three patients (25%) required further de-escalation to maintain tolerability at 80 mg.

Remarkably, only one patient exhibited tolerance to the 160 mg dosage but developed pneumonitis thereafter. In terms of 5-FU dosages, the majority, ten patients (83%) precisely, received a 400 mg/m^2 bolus and a continuous infusion of 2400 mg/m^2 over a 46 h duration, starting from the first day of therapy. However, four patients (33%) required a 20% dose reduction in 5-FU to enhance tolerability, and one patient received treatment in an external medical facility.

Overall, seven (58%) out of twelve patients had a good tolerance to treatment, defined as completing at least two cycles without grade 3/4 toxicities requiring discontinuation or hospitalization. These included Patients #2, #4, and #8–#12, who maintained regorafenib at 120 mg without severe adverse events, and Patient #1, who required dose reduction to 80 mg but continued therapy.

The least favorable response was in patient #5, who initially started regorafenib at a dose of 80 mg, with subsequent escalation to 120 mg. However, the patient developed grade 3 HFS. This adverse event necessitated the transition to an alternative therapeutic approach and a modified treatment plan.

In the case of patient #7, the initiation of regorafenib as monotherapy preceded the subsequent addition of 5-FU. Unfortunately, this patient experienced regorafenib-related pneumonitis, which entailed home oxygen support and discontinuation of the combined regimen. The patient experienced improvement with post-treatment care which included a course of steroids with a tapering regimen.

During treatment, three patients (25%) encountered mild adverse events that required adjustments to regorafenib dosages. For instance, patients #1 and #6 exhibited grade 2 HFS and grade 1–2 mucositis, respectively, prompting a modification of the regorafenib dosage from the initial target dose of 120 mg to 80 mg daily. Patient #3, with a pre-existing history of HTN, experienced hypertensive urgency marked by a systolic blood pressure exceeding 180 mmHg. In response, adjustments were made to the patient's antihypertensive medications, resulting in the resolution of hypertensive urgency without recurrence. The remaining patients exhibited tolerance to the medication at the intended goal dose of 120 mg.

Additionally, gastrointestinal toxicities were closely monitored due to their prevalence with regorafenib and 5-FU. Mucositis (grade 1–2) occurred in patient #6, starting on day 7 of cycle 1 and resolving within 1 week after reducing regorafenib to 80 mg/day and using supportive oral treatment. Notably, no cases of diarrhea were reported in this cohort, potentially due to proactive use of antidiarrheal agents (e.g., loperamide) initiated at treatment onset or the lower starting dose of regorafenib (80–120 mg).

Adverse events, including HFS, hypertension, and pneumonitis, were managed with tailored interventions: topical steroids and dose reductions for HFS, antihypertensives for hypertension, and steroids with supplemental oxygen for pneumonitis, as detailed in Tables 1 and 3. These strategies, optimized for this off-label regimen, ensured a 75% continuation rate. A comprehensive summary of treatment doses and associated adverse events for all patients is available for review in Tables 1 and 3.

Table 3. Dosing and reported adverse events for patients receiving regorafenib and 5-FU.

Patient #	Regorafenib Dose at Initiation	Regorafenib Dose at Last Follow-Up or Discontinuation	5-FU Dose at Initiation (mg/m^2)	5-FU Dose at Last Follow-Up or Discontinuation (mg/m^2)	Adverse Events Reported
1	120 mg	80 mg	Day 1: 400 Day 2: 2400	Day 1: 320 Day 2: 1920	Grade 1–2 HFS
2	80 mg	120 mg	Received in an outside facility	Received in an outside facility.	Well-tolerated
3	120 mg	120 mg	Day 1: 400 Day 2: 2400	Day 1: 400 Day 2: 2400	Grade 3 HTN
4	120 mg	120 mg	Day 1: 400 Day 2: 2400	Day 1: 400 Day 2: 2400	Well-tolerated
5	120 mg	80 mg	Day 1: 400 Day 2: 2400	Day 1: 320 Day 2: 1920	Grade 3 HFS
6	120 mg	80 mg	Day 1: 400 Day 2: 2400	Day 1: 320 Day 2: 1920	Grade 1–2 mucositis managed with regorafenib dose reduction to 80 mg and oral rinses, resolved in 7 days.

Table 3. *Cont.*

Patient #	Regorafenib Dose at Initiation	Regorafenib Dose at Last Follow-Up or Discontinuation	5-FU Dose at Initiation (mg/m^2)	5-FU Dose at Last Follow-Up or Discontinuation (mg/m^2)	Adverse Events Reported
7	160 mg	160 mg	Day 1: 320 Day 2: 1920	Day 1: 320 Day 2: 1920	Grade 3 pneumonitis
8	80 mg	120 mg	Day 1: 400 Day 2: 2400	Day 1: 400 Day 2: 2400	Well-tolerated
9	80 mg	120 mg	Day 1: 400 Day 2: 2400	Day 1: 400 Day 2: 2400	Well-tolerated
10	80 mg	120 mg	Day 1: 400 Day 2: 2400	Day 1: 400 Day 2: 2400	Well-tolerated
11	80 mg	120 mg	Day 1: 400 Day 2: 2400	Day 1: 400 Day 2: 2400	Well-tolerated
12	80 mg	120 mg	Day 1: 400 Day 2: 2400	Day 1: 400 Day 2: 2400	Well-tolerated

4. Discussion

Certainly, regorafenib offers promising outcomes for patients with refractory mCRC who have had disease progression despite all approved standard therapies. Clinical data indicates that regorafenib tends to yield better results in patients with good PS; therefore, it should be incorporated into the management of mCRC before patients become too frail and begin to experience a rapid decrease in PS. In our study, we highlight that using a combination of regorafenib plus 5-FU could be an optimal option for patients with refractory mCRC; through our investigation of treatment efficacy and safety, we provide insights into the potential benefits of this combination therapy, offering a valuable option for patients facing this challenging condition.

A study conducted by Fabien Calcagno et al. at Franche-Comté Cancer Hospitals included 29 patients who were treated with regorafenib monotherapy for mCRC, with doses ranging from 80 to 120 mg. In the CRC retrospective study, the median OS was six months and the median PFS was not assessed [13].

In our study, one patient (8.3%) achieved a PR, five patients (41.6%) experienced SD, and six patients (50%) had progression of disease. The median PFS was 96 days, (3.15) months. The 10-month OS was 61% (95% Confidence Interval (CI):25–84%) and the 12-month OS was 50% (95% CI: 18.8–75.3%). These findings indicate improved outcomes compared to previous research findings, suggesting that combining regorafenib and 5-FU leads to better results than using regorafenib as a monotherapy.

In our study, we observed impressive outcomes in two patients with mCRC who had previously undergone multiple lines of chemotherapy. The first patient was a 48-year-old male who was treated with an initial regorafenib dose of 80 mg and escalated to 120 mg. The patient was able to receive 12 cycles of regorafenib plus 5-FU therapy, with five prior chemotherapies consisting of capecitabine + oxaliplatin, FOLFIRI + bevacizumab, FOLFOX, trifluridine/tipiracil + bevacizumab, and pembrolizumab monotherapy. The patient began regorafenib as a sixth-line chemotherapy after undergoing sigmoid colectomy. MRI scan results showed no signs of tumor recurrence, indicating PR with stable liver metastasis until he had progression of disease. The other patient was a 59-year-old man with multiple lung and liver metastases who received seven prior chemotherapies and was initiated on regorafenib and 5-FU as the eighth line of therapy. Following the administration of two courses of regorafenib, the patient showed a decrease in CEA levels from 382 to 291 ng/mL, reached 177 ng/mL after two courses of therapy, and eventually experienced disease progression. This observation suggests a potential therapeutic benefit of regorafenib plus 5-FU therapy in controlling disease progression and reducing tumor burden in heavily pretreated patients with refractory mCRC.

Similarly, in another study conducted by Marks et al., two patients with mCRC who were resistant to first-line FOLFOX and second-line FOLFIRI were administered regorafenib in conjunction with either capecitabine or 5-FU. In both cases, despite prior treatment failures, the patients demonstrated evidence of disease control. One patient-maintained SD for at least one month with regorafenib and capecitabine therapy, while the other achieved at least two months of SD before starting to accumulate new metastatic foci. Additionally, in vitro studies revealed synergistic effects when regorafenib was combined with 5-FU, suggesting a potential mechanistic basis for the observed clinical responses. Comparing the outcomes of our study with those reported by Marks et al., we observe similarities in patient characteristics and treatment responses. Both studies highlight the potential efficacy of regorafenib plus 5-FU in refractory mCRC. The observed disease control and prolonged survival in some patients underscore the importance of exploring novel treatment combinations in this challenging patient population [14].

In our study, which investigates the combination therapy of regorafenib plus 5-FU in mCRC patients, we found that most adverse events associated with the regimen were of low severity, indicating a manageable safety profile within our patient cohort. However, a notable proportion of three patients (25%) required adjustments to the regorafenib dosage based on clinical considerations, with two patients discontinuing treatment due to intolerable adverse effects.

Our findings also revealed variations in the initial doses chosen in clinical practice, with 80 mg (n = 50%) being the most selected dose, followed by 120 mg (n = 41.6%) and 160 mg (n = 8.3%). Notably, a direct relationship was observed between the initial dose and the frequency of Grade 3 or 4 adverse events; as the initial dose increases, a higher frequency of Grade 3/4 adverse events was observed (0% for 80 mg, 8.3% for 120 mg, and 8.3% for 160 mg). Interestingly, these observations regarding dose-dependent toxicity align with recent findings from the ReDOS study, which demonstrated that a strategy of weekly dose escalation of regorafenib from 80 mg to 160 mg/day was non-inferior to a starting dose of 160 mg/day for survival outcomes. This suggests that a gradual dose-escalation approach may offer comparable efficacy while potentially reducing the incidence of severe adverse events [15]. The absence of diarrhea in this cohort contrasts with studies like Zanwar et al. [16], where 65% of patients experienced grade 3/4 toxicities, including diarrhea. This may reflect our use of lower initial regorafenib doses (80–120 mg) and proactive supportive care, such as loperamide. The single case of mucositis was effectively managed with dose reduction and oral rinses, highlighting the importance of early intervention to maintain treatment continuity.

However, our study's findings diverge significantly from those reported by Zanwar et al. [16] in a study conducted at a tertiary cancer center in India. Their investigations involved 23 patients treated with regorafenib; dose reduction was required for 86.9% of patients. Thirteen patients were initiated at a lower dose of 120 mg initially due to the poor tolerance of the 160 mg dose observed in the first ten patients. The occurrence of Grade 3/4 drug-related adverse events was notably higher, with at least one Grade 3/4 toxicity noted in 65% of cases studied [16].

This comparison highlights differences in treatment outcomes and adverse event profiles between our study and the study by Zanwar et al. [16], underscoring the importance of considering regional and patient-specific factors when determining treatment approaches for mCRC patients receiving regorafenib therapy. Overall, our study underlines the importance of personalized dosing and vigilant monitoring of adverse events in clinical practice, aiming to maximize treatment efficacy while minimizing treatment-related toxicity.

Further collaborative research efforts are needed to elucidate the optimal dosing strategies and improve outcomes for patients with CRC undergoing regorafenib plus 5-FU therapy. Hence, it is imperative to closely monitor patients during the initial course of regorafenib treatment to proficiently address and mitigate any potential treatment-related adverse effects.

Furthermore, our study delved into the examination of patients' status during their initial follow-up visit after the commencement of regorafenib plus 5-FU therapy. Conducting vigilant surveillance at these junctures facilitates the early identification of any adverse effects associated with the treatment, thus allowing for timely intervention before any exacerbation in severity. Prior clinical investigations have also underscored the favorable tolerability and moderate antitumor efficacy of an initial regorafenib dose ranging from 80 mg to 120 mg, particularly when employed as salvage therapy for mCRC [17].

The promising 50% DCR and 12-month OS in this rare cohort directly supported funding for a Phase II trial at Houston Methodist Neal Cancer Center (NCT06887218, HMCC-GI24-001), approved in February 2025, with a planned enrollment of 56 patients. Despite its small sample size and retrospective design, this study's success as a foundational step validates its role in driving prospective research. The constrained sample size reflects the off-label use of the treatment in refractory mCRC, limiting eligibility to a rare and highly specific patient subgroup. To confirm these findings and optimize treatment sequencing and combination strategies for refractory mCRC, larger prospective studies are essential.

5. Conclusions

The findings of our study suggest that regorafenib and 5-FU combination therapy is a possible treatment option in mCRC patients, especially when considering systemic therapy beyond the third line. This regimen warrants proactive evaluation against other salvage therapies. Further prospective studies with larger cohorts are essential to confirm these findings and optimize the efficacy and safety of this treatment for refractory mCRC patients.

Author Contributions: For the development of this study titled "Efficacy and Safety of Regorafenib Plus 5-Fluorouracil Combination Therapy for Refractory Metastatic Colorectal Carcinoma Beyond Third Line Treatment: An Institutional Experience", the contributions were as follows: M.A. and A.E. conceptualized and designed this institutional experience, with A.E. also providing administrative support. M.A., A.E., E.A.-N., B.K., G.U., W.A., K.A. and V.B.A. were responsible for providing study materials and patient-related data. The collection and assembly of data were handled by B.K. and E.A.-N., while B.K. led the analysis and interpretation efforts. All authors—M.A., A.E., E.A.-N., B.K., G.U., W.A., K.A. and V.B.A.—actively participated in writing the manuscript for this institutional experience and gave their final approval to the completed document. All authors have read and agreed to the published version of the manuscript.

Funding: This research received no external funding.

Informed Consent Statement: The authors are accountable for all aspects of the work in ensuring that questions related to the accuracy or integrity of any part of the work are appropriately investigated and resolved. The study was conducted in accordance with the Declaration of Helsinki (as revised in 2013). The study was approved by the Institutional Review Board of Houston Methodist Hospital (No. PRO00035386). Informed consent was waived due to the use of de-identified data collected as part of routine clinical care.

Data Availability Statement: Data are contained within the article.

Acknowledgments: We would like to express our deepest appreciation and gratitude to The Cockrell Center for Advanced Therapeutics, the William and Ella Owens Medical Research Foundation, and the Houston Methodist Hospital Foundation for their support.

Conflicts of Interest: The authors declare no conflict of interest.

References

1. Navarro, M.; Nicolas, A.; Ferrandez, A.; Lanas, A. Colorectal cancer population screening programs worldwide in 2016: An update. *World J. Gastroenterol.* **2016**, *23*, 3632. [CrossRef] [PubMed]
2. Siegel, R.L.; Wagle, N.S.; Cercek, A.; Smith, R.A.; Jemal, A. Colorectal cancer statistics, 2023. *CA Cancer J. Clin.* **2023**, *73*, 233–254. [CrossRef] [PubMed]
3. Biller, L.H.; Schrag, D. Diagnosis and Treatment of Metastatic Colorectal Cancer. *JAMA* **2021**, *325*, 669–685. [CrossRef] [PubMed]
4. Edwards, M.S.; Chadda, S.D.; Zhao, Z.; Barber, B.L.; Sykes, D.P. A systematic review of treatment guidelines for metastatic colorectal cancer. *Color. Dis.* **2012**, *14*, e31–e47. [CrossRef] [PubMed]
5. Haque, E.; Muhsen, I.N.; Esmail, A.; Umoru, G.; Mylavarapu, C.; Ajewole, V.B.; Abdelrahim, M. Case report: Efficacy and safety of regorafenib plus fluorouracil combination therapy in the treatment of refractory metastatic colorectal cancer. *Front. Oncol.* **2022**, *12*, 992455. [CrossRef]
6. Strumberg, D.; Scheulen, M.E.; Schultheis, B.; Richly, H.; Frost, A.; Büchert, M.; Christensen, O.; Jeffers, M.; Heinig, R.; Boix, O.; et al. Regorafenib (BAY 73-4506) in advanced colorectal cancer: A phase I study. *Br. J. Cancer* **2012**, *106*, 1722–1727. [CrossRef] [PubMed]
7. Cho, S.M.; Esmail, A.; Abdelrahim, M. Triple-regimen of vemurafenib, irinotecan, and cetuximab for the treatment of BRAFV600E-mutant CRC: A case report and review. *Front. Pharmacol.* **2021**, *12*, 795381. [CrossRef] [PubMed]
8. Sethy, C.; Kundu, C.N. 5-Fluorouracil (5-FU) resistance and the new strategy to enhance the sensitivity against cancer: Implication of DNA repair inhibition. *Biomed. Pharmacother.* **2021**, *137*, 111285. [CrossRef] [PubMed]
9. García-Alfonso, P.; Martín, A.J.M.; Morán, L.O.; Alsar, J.S.; Pérez-Solero, G.T.; Codesido, M.B.; Ferrandiz, P.A.C.; Cicala, S.G. Oral drugs in the treatment of metastatic colorectal cancer. *Ther. Adv. Med. Oncol.* **2021**, *13*, 17588359211009001. [CrossRef] [PubMed]
10. Crona, D.J.; Keisler, M.D.; Walko, C.M. Regorafenib:A Novel Multitargeted Tyrosine Kinase Inhibitor for Colorectal Cancer and Gastrointestinal Stromal Tumors. *Ann. Pharmacother.* **2013**, *47*, 1685–1696. [CrossRef] [PubMed]
11. Cousin, S.; Cantarel, C.; Guegan, J.-P.; Gomez-Roca, C.; Metges, J.-P.; Adenis, A.; Pernot, S.; Bellera, C.; Kind, M.; Auzanneau, C.; et al. Regorafenib-Avelumab Combination in Patients with Microsatellite Stable Colorectal Cancer (REGOMUNE): A Single-arm, Open-label, Phase II Trial. *Clin. Cancer Res.* **2021**, *27*, 2139–2147. [CrossRef] [PubMed]
12. Yang, K.; Han, L.; Wu, S.; Qu, X.; Li, Q.; Zhao, C.; Zhou, J.; Jin, X.; Wang, Y.; Yan, D.; et al. Real-world outcomes of regorafenib combined with immune checkpoint inhibitors in patients with advanced or metastatic microsatellite stable colorectal cancer: A multicenter study. *Cancer Immunol. Immunother.* **2022**, *71*, 1443–1451. [CrossRef]
13. Calcagno, F.; Lenoble, S.; Lakkis, Z.; Nguyen, T.; Limat, S.; Borg, C.; Jary, M.; Kim, S.; Nerich, V. Efficacy, Safety and Cost of Regorafenib in Patients with Metastatic Colorectal Cancer in French Clinical Practice. *Clin. Med. Insights Oncol.* **2016**, *10*, 59–66. [CrossRef] [PubMed]
14. Marks, E.I.; Tan, C.; Zhang, J.; Zhou, L.; Yang, Z.; Scicchitano, A.; El-Deiry, W.S. El-Deiry, Regorafenib with a fluoropyrimidine for metastatic colorectal cancer after progression on multiple 5-FU-containing combination therapies and regorafenib monotherapy. *Cancer Biol. Ther.* **2016**, *16*, 1710–1719. [CrossRef] [PubMed]
15. Bekaii-Saab, T.; Ou, F.; Anderson, D.; Ahn, D.; Boland, P.; Ciombor, K.; Jacobs, N.; Desnoyers, R.; Cleary, J.; Meyers, J.; et al. Regorafenib dose optimization study (ReDOS): Randomized phase II trial to evaluate dosing strategies for regorafenib in refractory metastatic colorectal cancer (mCRC) (a)over-cap(sic) An ACCRU Network study. *Ann. Oncol.* **2018**, *29* (Suppl. S5), v105. [CrossRef]
16. Zanwar, S.; Ostwal, V.; Gupta, S.; Sirohi, B.; Toshniwal, A.; Shetty, N.; Banavali, S. Toxicity and early outcomes of regorafenib in multiply pre-treated metastatic colorectal adenocarcinoma-experience from a tertiary cancer centre in India. *Ann. Transl. Med.* **2016**, *4*, 74. [PubMed]
17. Yoon, S.E.; Lee, S.J.; Lee, J.; Park, S.H.; Park, J.O.; Lim, H.Y.; Kang, W.K.; Park, Y.S.; Kim, S.T. The use of regorafenib for patients with refractory metastatic colorectal cancer in clinical practice. *OncoTargets Ther.* **2018**, *12*, 225–231. [CrossRef] [PubMed]

MDPI AG
Grosspeteranlage 5
4052 Basel
Switzerland
Tel.: +41 61 683 77 34

Biomedicines Editorial Office
E-mail: biomedicines@mdpi.com
www.mdpi.com/journal/biomedicines